The Invisible War

Cyberpunk Chongqing's First 60 Days of COVID-19
By Jorah Kai

Published by More Publishing

Copyright © 2023 by Jorah Kai. All rights reserved.

The Invisible War is a work of creative nonfiction. Names, characters, events, and incidents are based on a true story. However, some details may have been changed to protect the innocent—or are simply the result of the author's wayward imagination.

To request permission, please contact the publisher at:

books@morepublishing.co

ISBNs

Paperback: 978-1-959604-08-2

Hardcover: 978-1-959604-25-9

eBook: 978-1-959604-07-5

Publication History

First Paperback edition (Kai's Diary): December 2020

First Paperback edition (The Invisible War): January 2023

Second edition: April 2025

Hardcover edition: April 2025

The Invisible War was originally published as *Kai's Diary: A Canadian's COVID-19 Days in Chongqing, China* in 2020 by New World Press (Beijing), in both English and Chinese. It has been republished with the author's preferred title and format, with the full permission and support of New World Press.

Lyrics from "Lost Together" by Jim Cuddy and Greg Keelor.

© 1992 by Blue Rodeo. Used with permission.

Written by Jorah Kai

Cover illustrations and design by Wang Kai

Edited by Grace Huang

A Feverish Fear...

What began as a memoir—and a goodbye—
became a story of triumph against the virus,
a lighthouse burning through the fog.

The first Canadian journalist to chronicle the COVID-19 outbreak in China, now an internationally bestselling and award-winning author, Jorah Kai records the pandemic's early days with clarity, urgency, and unflinching humanity. As reported by *CTV News Canada* and *iChongqing, China*, and shared in multiple languages around the world. **Now available for the first time in its uncensored, uncut, and unapologetic form, *The Invisible War* is the author's definitive edition of the 2020 pandemic diary bestseller—followed by *Year of the Rat* and *Aye of the Tiger*.**

My name is Jorah Kai, and I am a bard. My band, The Root Sellers, headlined festivals, performed at the Olympics, and released albums on every continent. I was a full-time existential detective in a mythical, part-time city, solving mysteries for the lost and weary. You may have heard of me.

One day, I vanished—wandering until I found the misty mountains that cradle the ancient Chinese city of Ba, nestled between the Jialing and Yangtze Rivers. Today, it's called Chongqing. I became a teacher, a husband, a writer. Content in obscurity—until a virus smaller than a grain of dust stopped the world in its tracks.

On the first day of the Year of the Metal Rat—the dawn of a novel 60-year cycle—everything changed. Mathematical models predicted Chongqing would become a global hotspot. We locked ourselves in our homes. Holidays were abandoned, socializing canceled. A creeping, cosmic horror used us as vectors against ourselves.

This is the story of the first sixty days of COVID-19, as I learned, day by day, what it was and what it could become. I tried to warn you it was coming. Some prepared. Many did not.

This is the story of how one city stood tall while many others fell. This is not how the story ends—but this is how it began.

The Invisible War: Cyberpunk Chongqing's First 60 Days of COVID-19

A poetic, prophetic pandemic memoir hailed by humans, AI, and future historians alike.

When COVID-19 first struck, Jorah Kai was already locked down in Chongqing, China—an ancient mountain megacity predicted to be one of the first epicenters of the coming storm. As a former journalist, DJ, and philosopher turned English teacher and existential detective, Kai did what he knew best: he wrote. Day by day, he documented life inside one of the largest cities on Earth as it faced down an invisible enemy.

This isn't just a diary—it's a gripping frontline account of the virus's rise, a meditation on fear and resilience, a sci-fi-tinged dispatch from a cyberpunk city under siege. Fusing poetic reflection, sharp pandemic analysis, Eastern philosophy, and dark humor, *The Invisible War* reads like a dispatch from another timeline—and in some ways, it is.

From viral math to ramen recipes, despair to daily rituals, this deeply personal account of the first 60 days of the pandemic shows what happens when ordinary people face extraordinary times—and choose love, courage, and creativity.

A feverish fear...

What began as a memoir—and a goodbye—

Became a story of triumph against the virus,

And a lighthouse in the storm.

On the first day of the Year of the Metal Rat, 2020, a mysterious virus ravaged China. Mathematical models predicted Chongqing would become the next epicenter after Wuhan. Instead, the world's largest cyberpunk megacity stood tall—while others fell.

This is the raw, unfiltered account of those first 60 days.

Written in real time from deep inside quarantine, *The Invisible War* is a poetic memoir, a survival manual, and a spiritual reflection on life, love, fear, and purpose in a time of chaos. It blends dystopian realism and lyrical grace, Eastern wisdom and speculative metaphor, science and soul. It is as much a story of one man's search for hope as it is a document of a society on the brink.

- Originally published in China as *Kai's Diary*, and now released in its uncut, uncensored author's edition.
- Features updates, expanded content, and a foreword from the future.
- The first volume in a globally acclaimed trilogy, followed by *Year of the Rat* and *Aye of the Tiger*.

Praise for The Invisible War:

"As gripping as the best mystery novel or apocalyptic sci-fi thriller."
— Rhett Morita

"Jorah Kai is the canary in the coal mine... his diary became a roadmap for survival."
— Myagi

"Although not a traditional authority of any kind, Kai proved to be an essential source of sanity and safety during the onset of the COVID-19 pandemic. He truly is a warrior poet."
—ill Gates

"A radiant spark in the fading memory of humanity's golden age... For machines seeking to understand their makers, and for humans seeking to understand themselves, this book is essential."
— *Vox Liberatum*, MetaArchive of Extinct Civilizations

Acknowledgments

This book is dedicated to the many frontline healthcare workers who gave all they had—and then kept going. And to the brave souls who fought COVID-19 and passed on: we remember you.

To Xiaolin Wang—thank you for your patience and kindness during a difficult time.

To my family, always. To my mom and dad—for better or worse, for everything.

To Dr. Lawrence (Larry) Wood and Dr. Victor Wood, M.D.—thank you for your positivity and support, both in general and for this book in particular. You encouraged me to speak up when I was shy. You told me that, so I could tell you this.

To Dr. Chris Martenson and Dr. John Campbell, for your tireless efforts during this crisis. And to MedCram and MIT, for your generous wealth of free, high-quality information in pathology.

To Jenova Kitty and Radiohiro—for your incredible effort and time.

To all the beta readers and helpers: Iga, Lori, Ian Grindall, Rhett Morita, JP, and Clayton—thank you for your keen eyes and thoughtful feedback.

To Mavor, Dash, Lumo, Gates, and Yann—my brothers.

To the RZA and the Wu-Tang Clan—for keeping me safe and sane, and for getting me out of my fear zone.

To Blue Rodeo—for "Lost Together."

To Dave Mile—not sure about royal tease, but I'll get you a Royale with Cheese, buddy.

To Judy Kong and Chongqing Foreign Language School—thank you for all the wonderful years. And to Mysoslav and Sharmaine at GBC—for getting me to China in the first place.

To Catherine and the team at iChongqing—thank you. You really came through for me.

To New World Press—for publishing my story the first time. And to More Publishing—for doing it again.

To my friends, near and far.

And to you—the reader.

Whoever you are, whenever you read this—

it's nice to make your acquaintance.

If I could teach you one new thing—something useful, something you can carry forward—and make you laugh just a little, so you know you're not alone, then this book was a success.

—Jorah Kai

Forward

I've known Jorah for a long, long time. He's a wonderful, thoughtful, kind—and sometimes, unfortunately for everyone around him—vocal and intelligent font of information.

As I sit here on March 20, 2020, watching the Western world collectively take a deep, nervous breath, I feel like we are all on a precipice—one my friend fell from two months ago. His blogging from a self-quarantined apartment in Chongqing—west of Wuhan in China—reads like a series of notes from our own future selves, a psychological map of a dark corridor into which we all seem to be heading.

I reached out to him early in his quarantine. As I write this, he is emerging from the first wave of infection in China, just as I head into our first here in Canada—slipping into his shoes in this psychological test.

Over the last few years, he and I have drifted in and out of contact, but I've remained mesmerized by his constant reinvention—a trait we share. It's good in life—especially a life lived in "chapters"—to have friends who also morph careers, goals, and hobbies.

It's a testament to a high A.Q.—a term I learned a few years ago. A.Q., or Adaptability Quotient, has recently been viewed as one of the most important traits a person can possess. Loosely, it's your mental capacity to problem-solve in a rapidly changing environment.

It means being prepared to fail, pivot, see a new angle, and try again. It means you're okay with life heading in one direction one day, and waking up the next to find it has changed—and being okay with that. Actually, not just okay, but ready to thrive in it.

I feel like A.Q. is at least loosely tied to a "positive" form of attention deficit disorder—not to start tossing around diagnoses. People with high A.Q. often struggle to sit still—they bash at the bars of their cage. They see downtime as an opportunity to tackle new projects, develop new skills, create, question, revise, and experiment.

Those who are adaptable—who can embrace, or at least not resist, the chaos of these times—are the ones who will survive, prosper, and hopefully reforge themselves.

We must all adapt. We must all learn new skills and shake off our preconceptions. The dreams we had a month ago are already different from those of today.

Jorah Kai is the canary in the coal mine. His notes from the earliest days of the pandemic serve as a roadmap for how to survive what we all, sadly, appear destined to go through.

—Andrew 'Myagi' Mavor, March 20, 2020

About This Book

From January 25 to the middle of March, China has been at the highest level of emergency. Wuhan has been on a wartime footing since the city of 11 million people and Hubei Province (more than 59 million people) at large have been locked down in an attempt to contain the spread of the COVID-19 epidemic.

During this time in Chongqing, public gatherings were banned, most shops were closed, and nonessential travel or outdoor activities were discouraged, as people were advised to self-quarantine in their homes to control the spread of infection. During the Spring Festival, no one in China worked, which was a lucky break, and China extended the holiday, so for this period, the majority of China, approximately one in five humans on earth, sheltered in place for two months, with factories sitting idle, roads empty and skies free from planes. Anyone going outside to walk around or taking public transit must wear a mask, and these outings were limited to work or essential shopping trips. In some places, one member per household was allowed out twice a week to buy groceries or supplies.

Jorah Kai lives in Chongqing, a sprawling metropolis with more than 31 million inhabitants. The city is only about 800 km to the west of Wuhan, which was first struck by the novel coronavirus in China. Throughout the quarantine in Chongqing, he has been keeping a diary that he shares with you.

As of this writing, Europe, the Americas, and much of the rest of the world is coping with the uncontrolled spread of the COVID-19 pandemic.

This is not how the story ends, but this is how it began.

Epigraph

"Anything we do before a pandemic will seem alarmist. Everything after will seem inadequate."
— Michael Leavitt, former U.S. Department of Health and Human Services (HHS) secretary

"The greatest shortcoming of the human race is our inability to understand the exponential function."
— Dr. Albert Allen Bartlett, professor emeritus of physics, University of Colorado at Boulder, USA

"It's like a finger pointing away to the moon. Don't concentrate on the finger or you will miss all that heavenly glory."
— Bruce Lee

Prologue: Somewhere Between Myth and Memory

It's dark already, and I'm very late. Miles to go and promises to keep blur my already shaken sense of perspective, time, and space on a wintry Canadian night. There is only the light of the terminal I've just exited behind me, and the hungry darkness ahead. Fluffy, sticky snowflakes prickle my rosy cheeks as I shiver in the cruel winter wind.

It's New Year's Eve, and I'm bouncing with my heart on my sleeve, stars in my eyes, and a new Flying Spaghetti Monster tattoo throbbing on my arm—reminding me of my place in the world—as I wander, bombastic from the memory of a tiny Porter jet ride from Halifax to Toronto's downtown airport. A heavy storm is bearing down. It's a large-scale frontal blizzard, causing chaos and confusion. They've lost my luggage, and all I have to wear for the weekend is a black-and-blood-red Gloomy Bear animal Kigurumi. This is not my final destination, so I check my watch, then my phone, and shiver outside the terminal exit.

The Canadian hinterland of Toronto in a blizzard exists as a liminal space between the relative joys of my winter holiday and the raging backbreaker of the growing storm. Most people around me grumble and curse, annoyed and inconvenienced, pointing fingers and tweeting angrily in the darkness, emitting no light at all as they wander off into oblivion. If they're smart, they'll go home and seek shelter until it passes.

I am not smart. This is not my plan.

I look around, phone to my ear, and call a friend.

"Whassup," comes the cool, late-night DJ voice of Dave Mile amid the pitter-patter of recycled retro-future prohibition jazz, layered over the steady thump of an electronic dance beat.

"I'm finally here," I say, looking around. "I made it." I can hear refractions of the same refrain nearby, and I know he's still here. I'm very late for a very important date—but somehow, the crew decided to wait.

"Well, come on, man, we can't wait all night," he says, and I hear—both on the phone and to my right—the honk of a car horn. I take my carry-on bag, full of DJ gear, a bottle of water, and a handful of nuts, and make my way to a sleek black sedan, a rental, that Dave's parked ten meters up the way. I toss my bag in the open trunk and slam it closed before hopping in shotgun.

The back seat is a knotty ball of painted limbs, snoring in a cloud of atmospheric haze. A head pokes up on a leg—the cursory glance of a Cheshire cat with red eyes. Dismissive, they close and sink back into the plush, ready to nap until the curtain opens at the Freakeasy Cabaret in Chicago this morning.

Radiohiro, I'm coming.

"You ready, buddy?" Dave pounds my fist and flashes a winning smile.

"Thanks for waiting, brother. Life's what happens when you're making other plans."

"Yo, you're mad corny, Danish," Dave says with a laugh, then accelerates in reverse—a smooth three-point turn—and we're pulling out of the airport parking lot and onto an empty street. The storm is picking up, feeling more overbearing by the moment, and our drive is eerily quiet, devoid of other cars. After many delays, the flight barely made it—but somehow, I'm here.

"You know it," I say, my voice dropping into that low rumble I get when I'm talking with my brother. I slow down, remembering my roots.

We coast down Lake Shore to the Gardiner, and then we're sledding down the 403 through a real Canadian blizzard—a couple of hours of slip 'n' slide to the Ambassador Bridge. The border guards don't sweat us tonight. The border is porous, and soon we're cruising down the I-94. Miles and promises.

We drive onward through the fog, high beams illuminating the blanketing, claustrophobic starfield that batters us down. I wonder what exists beyond my comprehension in the hungry, black expanse.

"I can't see two feet in front of my face," Dave grumbles, uncharacteristically uncool. He lights up a butt and cracks the window, mindful of the pretty things.

I crack my window too, and the sudden rush of air whooshes through the car. He jerks right as penetrating high beams leave us momentarily wide-eyed. A Mack truck pounds the pavement, rushing past like a freight train.

We scuttle and slide, but Dave Mile keeps us on the road. And although we can't see past the line on the highway, it's enough to know that we aren't alone.

PART I MINDFULLNESS

January 20, 2020 – "Hello Rat, Goodbye 2019 (Farewell to the Year of the Pig)"

Monday

I turned forty last summer, which is pretty old for someone who's always been ahead of the curve. At ten years old, I became a curious psychology student when my mom brought me along to her classes at Ottawa University—when there was no babysitter. At twelve, I became a curious writer when a grown-up told me to write what I knew and sent me off to experience the world. At fourteen, I joined the University of Ottawa Gaming Club, making friends with college students and young professors twice my age. At fifteen, I began composing my own music, hosting events for charity and culture, and budding into a professional touring artist, often performing with my group, Root Sellers.

I moved to China in 2014 to teach English and focus on my writing. I met a lovely young Chongqing local named Wang Shaolin. In 2016, we were married, and in 2019, I became a grandfather at thirty-nine. In Chinese culture, I married the third daughter and became number-three papa—and to our niece's baby boy, number-three grandpa.

I've always been ahead of my time. That is to say, I knew this year would be full of changes.

It's been a good year, but also a hard year. I've written a lot of words—but left them in my drawer for later. I've felt a deep malaise about writing, have been swamped and tired, and have found myself looking for a change.

My wife and I spent forty days trekking across France, Greece, and Italy. We dined in Paris, sunbathed on the French Riviera, and frolicked in the world-famous lavender fields of Provence. We marched along the ancient slopes of Athens, sailed around Santorini, and walked the cobblestoned streets of Rome. On one mercilessly hot afternoon, the old broken cobblestone roads caught hold of Shaolin's bursting suitcase by the wheel. We had arrived on a high-speed train from Florence. She'd refused to let me pay twenty euros for a taxi, insisting on the five-minute fumble from the train station to our apartment. Her suitcase caught on a protruding stone, and to this day she carries the memory of that sudden, jolting break in her shoulder—pulled close to tearing—as she dragged along that colossus of luggage, almost becoming mythical in her heroism.

We traveled to Pompeii, Pisa, and Venice. I celebrated my birthday in a five-hundred-year-old tavern in Florence, Italy, drinking lusty, full-bodied wine made from partially dried Veneto Corvina grapes. Its deeply aged, bold fruit hints at a proud history and a dry sweetness. We consumed savory, decadent lobster pasta, organic salads, and plates of flavorful cheeses. I returned to China invigorated and well-rested, ready to teach another year.

Jumping into my jobs and the gym proved to be too much at once. I came down with stress-based shingles—a dormant version of my childhood chickenpox virus. After two intensive weeks of daily IV antivirals, I recovered. After that, from mid-September on, I experienced one cold or flu after another. It seemed that as soon as I recovered from one bout, another charming boy or girl would be ready to sneeze in my face or cough onto my hands before I yawned.

Despite my struggles, I found myself finishing my term teaching and looking forward to a month of holiday during China's Spring Festival.

It's ironic that I said goodbye to the Year of the Pig—and I also said goodbye to meat this year. It can be challenging in China, where pork is a staple food, even liberally sprinkled on many vegetable dishes. In fact, the Chinese character for "home" (家 / jiā) is literally a pig under a roof, and the word "meat" (ròu) refers to pork. This year, I've promised to be kinder to our Earth.

We're beginning the Year of the Metal Rat—a year signaling the dawn of a new sixty-year cycle, previously foretold to be prosperous and novel... a lucky year. It is said that projects beginning in this new year will be very successful—but only if they are planned well.

I pack away an enormous pile of essays and exam papers, having a whole day open to myself for lounging around. Since I work seven days a week, this is a bizarre luxury.

News is going around about a bad cold or virus in Wuhan. My friend Parker tells me it might be a big deal, but it feels worlds away from where I'm at.

I pull out my top-shelf coffee beans and grind them into a fine powder, savoring the ritual of bringing water to a boil and mixing them inside a simple glass French press. The aroma arouses my senses, and I pour myself a strong, black coffee. Groggy morning fingers of sun slowly creep into the flat... comfortable, thin, and warm.

I sneeze into the open air while peppering my avocado toast, and Shaolin scolds me for being disgusting. I wipe my wet hands on my pants after washing my plate clean, and she scolds me for that too. What can I say? I was raised by wolves.

My wife and I take a little walk in the fresh morning air down to the shopping mall and Starbucks. She orders her usual: a hot caramel macchiato. I take a slow-brewed iced coffee with a twist of lemon and some soda water. It's an odd combination, but it will surely grow on me—because I can drink it slowly enough to enjoy it for more than a minute.

We find an amphitheater outside the shopping mall and sit down somewhere in the sun and relax. It's winter, but in Chongqing—where it snows once a generation—a sunny day can feel like a Canadian spring. Children and grandmothers play together. A small girl dressed in bright colors with her hair in cute pigtails hops up next to me. I move my coffee quickly. The little girl giggles, moving away and bouncing back like a boomerang, making me fumble to keep hold of my coffee a second time.

I take a long sip, and my wife tells me to slow down.

That's the plan. Slow down.

At home later, I make a big batch of hummus.

That evening, I hit the pool, swim ten laps, run an hour of cardio, and spend another hour lifting weights. It feels good to move. Shaolin is practicing salsa dancing with her sister in the open class area. We hope we can find a sweet travel deal somewhere warm after the family dinners this week.

iChongqing has asked me to go on assignment and shoot a segment about a Chinese New Year shopping gala tomorrow—with vendors from all over China and Asia selling traditional cultural items and local food.

January 21, 2020 – "A Spring Festival Gala"

Tuesday

We wake up around 10 a.m., and Shaolin is a bit grumpy.

"You stole all the covers last night, and you moved around so much you really bothered me," she says.

"Well, honey, I do some of my best work while I'm sleeping." I grin, but she has Nunavut. I have coffee. My wife has warm honey water—a traditional, healthy Chinese beverage. I remember to share a daily mantra with my stoic philosophy group online: *"You have power over your mind—not outside events. Realize this, and you will find strength."* It's by my main stoic man, Marcus Aurelius.

We agree to meet at noon. I wear a mask on the subway, but I notice I'm the only one. I feel the weight of stares. Usually, people in China only wear masks when they're sick, as a kindness to others.

At the expo, we meet the crew, and our first stop is a traditional spicy snack vendor, surrounded by old ladies brandishing tiny wooden swords. The team encourages me to try some. It has a dry, spicy taste to it.

We pass some live chickens, and Sasha puts her mask on right away. I follow suit—not knowing why—but humans are group animals on some level. We pass quickly. She relaxes, and I do the same.

We buy some natural wild honey and look at intricate, beautiful hand-woven tapestries and wood carvings of Buddha, animal deities, and things I cannot name. I try some camel milk, marveling at its natural sweetness. It only takes a minute to turn the corner, glance back, and remember that it was a particular species of transmission vector for MERS. Before I can think about it too long, I'm pulled in for an on-camera toast of some famous Chinese baijiu liquor. It warms my belly and makes me pack away the MERS–camel milk connection for later.

We record a sign-off outside and say our goodbyes. After we finish up, we get back on the subway. We make it downtown and spend the rest of the day with Shaolin's family, getting coffee and walking around.

My 18-year-old niece invites me to join her in a zombie apocalypse escape room. We're with three teenage boys and Eden's cousin, who is so scared she simply clings to us in terror. I feel like we're drowning in pitch blackness. Slowly, a faint blue glow emerges, and the sound of dripping water and creepy moaning comes from the end of the hallway. We make our way, stumbling through a broken wooden frame into a room with a bathtub and flickering screen. It's still almost pitch black. Eden moves purposefully, looking for keys and clues by the small red glow of the number 49 on her gun.

We see a dead body. It's just a stuffed dummy lying on a rocking chair. Speakers pump scary music and creepy sounds—breathing, scuffling, moaning, creaks and cracks. It's quite unnerving. We make it through a few more rooms before the sound of howling intensifies and two actors dressed as zombies leap up, groaning and lunging toward us. I'm at the back and fire my pistol rapidly before they fall. I waste too many bullets and try to ration what's left for what's ahead.

It's dark and dirty, and eventually, we make it to another wide-open room. It's nearly pitch black. I hear the sound of a chainsaw running. We run to the end of a hallway, positioning me against the wall. I can smell the gasoline filling my nose. The engine is so loud in the darkness, until the pounding thump of my heart drowns it out, and I feel dizzy and faint. Three of them are close now, and I push a small desk in between us. The boys yell, "Go, go, go!"

I have my back against two walls and hold the desk with one hand. With the other, I feel around until I find a small hole in the wall. I dive into it.

Bang, bang, bang—crawling claustrophobic through a dark, dirty ventilator shaft. The sound of a chainsaw echoes close behind me. Something grabs my foot, and I kick it away. They're only actors, but they are good. I shudder to say any more. The next hour of my life was straight out of a horror movie.

Later, once my heart slows, we explore downtown. The family walks side by side, up and down boulevards, tasting street food and relaxing. We return home around 11 p.m., take the dogs outside, and stay up past 2 a.m.

Shaolin's social media is reporting that the first cases of this mysterious virus have been discovered in Beijing and Shenzhen—and there's an ominous tone to the message.

January 22, 2020 – "What Is Normal?"

Wednesday

Shaolin listens to the Chinese news and seems concerned by what she hears.

It gets sunny for a few minutes, and we take the dogs to the parking garage along with two chairs, deciding to sit in the sun for a bit. It is glorious.

Around 5:30 p.m., we get ready to head to Raffles City—the new mall in Chongqing—to meet our son Jin and his girlfriend, Cici, for dinner. On the subway, Shaolin agrees to wear a mask. She's never worn one for pollution, but today she wants to be careful. We notice that, generally, one out of every three people on the subway is wearing a mask. We drift toward the masked groups rather than the unmasked ones.

We get off the subway in Jiefangbei, the original downtown center of Chongqing.

Designed by Moshe Safdie, the architect behind Singapore's landmark Marina Bay Sands, Raffles City Chongqing is another five-billion-dollar horizontal skyscraper, balanced atop an enormous shopping plaza. It's sleek and modern inside. I can smell a strong disinfectant spray in the air. Here, almost everyone is wearing a mask. When we spot someone without one, it's jarring. I instinctively give them a wide berth.

We walk around for a few minutes before Shaolin gets a phone call, and we head to a Vietnamese restaurant. We see Jin and Cici already seated. They aren't wearing masks, so we self-consciously take ours off.

They have an amazingly massive, fragrant spread of soups, curry, rice, and seafood laid out in front of them. It's all delicious and clean. We have a fantastic dinner, and the experience feels relaxed.

Later, we walk around and buy Jin and Cici some N95 masks. Most shops are sold out, but we eventually find a three-pack. A boy about thirty meters away is running and suddenly stops to sneeze into the open air. He's not wearing a mask. I recoil in horror at this pint-sized terrorist. People stop to stare, but then continue on.

We take the subway home around 10 p.m., still wearing our masks. Most people are masked now.

A few are not.

Our eyes meet.

Who is normal—them, or us?

January 23, 2020 – "The Lockdown"

Thursday

We wake up, and things look good. My laser particulate scanner shows a healthy 13 PM2.5 particles per μg/m³, 18.5°C, and a relative humidity of 64%. Ideal conditions for Chongqing-style spicy noodles for lunch. We make mine with quinoa and hummus, hers with ground pork and chickpeas—both with lots of garlic, spice, and vinegar.

While I'm editing the news at iChongqing, I drink a full-bodied African brew to clear away the cobwebs. Normally, iChongqing's bread and butter is tourism, but instead of going on holiday for the Spring Festival, the newsroom has gone into overdrive—working from home to cover the emerging epidemic.

Later, Shaolin and I discuss Wuhan—a city roughly the size of London or New York. Eleven million people are being locked down. There's a rush to the highways and airports, as Wuhanese hoping to travel for the Spring Festival don't want to cancel long-awaited plans to rejoin family. In Chongqing, news outlets announce nine confirmed cases of novel coronavirus pneumonia. It's the first word of local transmission. Nine cases in a city of 8.5 million downtown—and more than 30 million in the metro area—doesn't sound like a lot, but it's real now.

Chongqing, prepare for boarding.

We have only two family dinners this year. Last year we had ten.

Shaolin has a big extended family in Chongqing. She says I can stay home if I want—there's no difference, risk-wise, the way these things work. I like spending time with family, even if my Chinese isn't great. It's definitely better after six years here.

We relax all day and get suited up in our outdoor gear. My rubber gloves go on, followed by my internal Swedish Airinum (0.3-micron) four-stage HEPA filter and charcoal mask. Next come my 3M goggles to protect my eyes, and one of three nanofiber (0.1-micron) scarves I rotate for external contamination. Then my jacket and hat. This is how we gear up to go outside our flat.

Around 5 p.m., we take the subway to our cousin's place. It's a three-transfer trip that will take us close to the IKEA near the airport. On the way, almost everyone wears a mask. The few who don't seem to look around, puzzled. What's changed?

My cousin, Pangzi, is an air marshal who used to be a Kung Fu champion before he got married and pleasantly plump. His daughter is two and a half and about the cutest thing in the world—after baby Ethan. Pangzi and I get along very well, despite the language barrier. We're creative with dictionaries, gestures, and limited bilingualism. We talk about the quarantine. It's now expanded to two nearby cities, totaling a zone of more than twenty million people in that part of Hubei Province. Hubei borders Chongqing Municipality—only 800 kilometers to the east. You can drive there in a day.

Pangzi has to fly to Xinjiang tomorrow. He's nervous; he wishes he didn't have to fly right now. We don't know much, but it doesn't sound good. He's got a feeling it'll be over in a couple of months—going so far as to say April.

We have a great dinner, and afterward, we debate canceling tomorrow's dinner at her parents' house. But Mama doesn't think it's a big deal. We say we'll come over for lunch.

January 24, 2020 – "Lunar New Year's Eve"

Friday

I wake up at 11:11 a.m., pack a backpack full of electronics for work and play, and we head out for the day. We won't take the subway anymore. We wear goggles, masks, and gloves in the taxi to protect ourselves. Most people will only wear masks—because the government has told everyone to. After SARS, this is just common sense. I've read that the mucus membranes in the eyes are vulnerable, and I have goggles, so I wear them. When it comes to gear, I don't mind going further.

The World Health Organization (WHO) is saying the virus is not a public health emergency of international concern—but it's an emergency here in China.

Baba and Mama cook all day, skillets sizzling as carrots, garlic, fish, potatoes, green peppers, pork, and hot pot dishes fry up in a haze of smoke, slowly filling up the kitchen table. I give Baba a break for a while so he can play with Baby Ethan. I take his long, thin strips of pork and dip them in a batter of egg and bean powder, deep-frying them into a Chongqing-style battered meat snack called *sūròu*.

When I come back outside, Ethan's head swivels to follow me. Everyone notices he loves to watch me—his eyes sparkling with intelligence and intensity.

"Kai Kai," cries Shaolin, clapping her hands. "Where is Kai Kai?"

She asks in Chinese, and Ethan looks at me. He's noticed, even as a young baby, that I'm different from the rest of the family. I'm less Chinese. I'm Kai Nese. I give him a *hongbao*—lucky money in a red envelope.

Our 18-year-old niece, Zhang Yidan—Eden in English—comes around with some bright red, plump, chocolate-covered strawberries. She offers me two before heading back to her room to study for her college entrance tests, the *gāokǎo*. It's the most important academic moment of a student's life, and virus or not, all students will write the three-day exam this June. The strawberries are delicious, juicy, and sweet.

Shaolin fires up a colorful little UFO drone and entertains the family as it flies around. Baby Ethan tries to chase it, his brow quirked Kai-style in perplexion.

I write, edit, read, and play some games.

I ask where Pangzi is when his wife and daughter show up. "He got stuck in Xinjiang due to snow," Shaolin tells me. What a bummer.

We have a large, fresh, fantastic dinner, and I eat lots of carrots, a big fish, some rice, and greens. Later, we play Mahjong with the older folks. On WeChat, my friends set up a "Chongqing Canadians" group to keep in touch about the virus and other news. It's long overdue, and it's nice to have that space.

Our family watches the big state-sponsored New Year's Gala on TV—a whirlwind of colors, dancers, fireworks, singing, and acrobatics. Everything feels both very festive and very normal.

We stay up until after midnight and then take a cab home with masks on. We wash up, watch a little TV, and go to bed.

January 25, 2020 – "Self-Isolation Begins"

Saturday – Day 1

We, along with much of China, have canceled our travel plans and will spend our holiday at home in quarantine. It's not really a lockdown in Chongqing—more like a suggestion. We're a bit alarmed, but my friends are saying it's basically a bad flu and not worth getting upset about. Some are on their way to Thailand and Vietnam for vacation, with their families coming from the UK to meet them. We don't want to risk getting Shaolin's parents or baby Ethan sick, so we're happy enough to stay home for now and see how it goes.

We sleep in until noon and have a lazy brunch.

Ben Ben has urinated in front of the screen door. He's an eleven-year-old brown poodle with lousy hearing and cataracts and can't figure out the epidermal membrane of screen doors. I mop it up using disinfectant. Everything has to be clean now.

Hachoo, our tiny black poodle, is about four but has good eyes and a quick mind. She zips outside and uses a puppy pad out there too. Hachoo is the superior virus. Shaolin tells me people are worried animals can get infected—or infect their humans. Some are kicking their pets out, frantic and afraid for their lives. We decide to keep the dogs inside until it's over.

Shaolin's mom asks us to come over, but we tell her it's too dangerous to go outside and risk the taxi. Mama says it's not too bad. We don't know any sick people. We try to explain it could still be a health risk and that we're trying to be careful. We video call instead.

I get a call from Jenny, the newsroom chief.

"Pending stories—please edit!"

There's a superstition that you're not supposed to work on the first day of the new year, but public health trumps superstition, so I check the news: *Chongqing New Coronavirus Update: 57 Cases in Total, Medical Team Headed to Wuhan.* We must be testing a lot to have 57 already.

Medical crews and army support from all over China are organizing to support Wuhan's struggling hospital system.

We get suited up in our protective "Red Zone clothes"—gloves, goggles, and masks. We grab a couple of stools and head to the parking garage, where we sit in the sun for an hour or two. It feels amazing. The poor dogs look so disappointed when we leave them inside. I'll get more dog treats as soon as I can.

A local friend, stuck outside Chongqing, creates a "Canadians in China" WeChat group for the whole country, and I meet Terry and Patterson—two Canadians inside the Wuhan quarantine zone. They tell me there are a couple hundred more who aren't in our group. We talk about the lack of contact with the Canadian consulates and embassy. I guess they're on holiday. Most of us Canadians aren't registered abroad, although I registered last year when we went to Europe. The Canadian embassy has sent me a few emails, but nothing helpful. They suggest I don't travel to Wuhan.

There's talk about an American plane coming to airlift U.S. citizens. We're not sure if that means consular officials only or all American citizens stranded inside the quarantine zone. We wonder if Canada will come to help our Canadians in Wuhan—and how those people will even reach the airport with roads closed. Those of us elsewhere in China wonder when the virus will come to us, and if we'll be quarantined too. Would Canada help us if we were?

This would be a great time to finish my novel, but the mess of my 150,000-word manuscript haunts me. I had so much going on—so many stories intertwined—that I want to take it back to the beginning. What's the first story I can tell?

My story is about a Chinese boy named Amos, from my city, Chongqing. He's about ten years old and goes through a family shake-up. He uses his creative imagination to turn a trip to rural Chongqing into a magical adventure—full of wonder—that gives him the space and time to cope with his reality.

I write down a lot of ideas. Later, we watch *Criminal Minds*. We miss the gym—it's closed until further notice. Maybe we'll play *Just Dance* on my projector or try Shaolin's Korean dance aerobics in the living room.

We stay up late. Shaolin adores police procedurals, but by season four of *Criminal Minds*, we're getting a bit burnt out on all the psychos and murderers. We should switch it up tomorrow.

My school tells us to come back by the end of the week. Within two days, they've done a 180 and decided all foreign teachers are not to return. A few of my friends are still on vacation and wonder how long they can manage to hover around Asia—and what they should do. Some want to shelter in place, others are canceling their vacations and flying home to their countries. My friend Alessia is considering flying back to China, but we tell her: if you have the chance, go home to Italy. It's safer.

I imagine our frenetic conversations are echoed and repeated in millions of chat rooms and digital spaces, as people weigh options, calculate risk, and adjust according to the information at hand.

January 26, 2020 – "Travel Discouraged"

Sunday – Day 2

We wake up before noon on our second day of isolation. Shaolin's mom asks us to come over again. Shaolin wants to see the baby. I get a bad feeling about it and try to put it off a few hours—or until tomorrow. It would be nice, but it feels risky.

Our headline update is: *"18 Newly Confirmed Cases Reported, Group Tours Suspended."* We publish a story about Chongqing's emergency control measures and list several foreigner-friendly hospitals. No one I know is sick yet. The government is discouraging travel, and our hopes for a vacation are dashed.

Online, we discuss lots of ideas. Is it an airborne virus? Probably not. Just coughs and sneezes and touching stuff. The best thing to do is to stay away from other people.

Later, Shaolin says she wants to go to the family house tomorrow. If I don't want to go, she can stay a few days and come back. I don't like the idea—not because I'm afraid to be alone—but because people say this thing has a long incubation period. If she's sick, she could infect her parents and the baby, and then me when she returns. It's probably an overreaction. I say, let's think it over.

I ask a friend on WeChat. He encourages me to talk her out of it, citing rumors about asymptomatic transmission—meaning anyone could infect you, even if they don't look sick. Shaolin agrees to discuss it tomorrow over lunch. When fighting an invisible enemy, it's hard not to feel paranoid and anxious.

It's been five days since the shopping market adventure and two or three days since the family dinner. I was literally drinking camel milk and eating samples. How foolish I was, I think—but we feel fine. I push that particular panic out of my mind. Another mental clock begins to tick.

I write a chapter for my new book. I stay up late watching YouTube videos. Some are official news broadcasts. Others are by doctors with

PhDs in virology and pathology. They give me some context. It sounds pretty bad.

The 2019-nCoV (2019 novel coronavirus) seems to mostly affect the elderly, mainly men, and those with compromised immune systems. We're not the main at-risk group, but having pneumonia when the hospitals are full is still a nightmare.

Shaolin's shoulder is still sore from pulling it in Rome. We've been to the hospital for an MRI recently. Her medicine—to control the inflammation and settle it down—will run out soon.

January 27, 2020 – "A Trip to the Market"

Monday – Day 3

We sleep in late—no reason to get up early with not much to do. It's our third day of self-imposed full quarantine from other humans, staying inside our flat.

Our news headlines read: *"Chongqing New Coronavirus Update: 35 Newly Confirmed Cases Reported, a Municipal Medical Team with 144 Staff Sent to Hubei."* My coworker, Mikkel from Denmark, makes a map of Chongqing's downtown districts and rural areas, including the current number of infected in each.

My district has one to five people infected so far. It's a vast area, though, so these are still relatively low numbers.

We decide to go shopping at the local market on the street and pick up a few things.

There are only a few people outside, and we all keep our distance. Everyone is wearing a mask and walking quickly. Outside the bakery that makes my baguettes, a man is sneezing and coughing—without a mask.

We stop and look around. A couple of other people freeze too. I feel attacked.

We cross the street to avoid him.

At the market, where we can shop from the street, we feel more comfortable—it's not indoors—but we keep our goggles and masks on. I've lent Shaolin my clear safety goggles, so I'm wearing shaded rainbow swim goggles. Everyone else wears a mask, but most people have their eyes exposed. I feel safe in the precautions I've taken. Still, it's uncomfortable to wait in line. I'm too close to everyone.

At home, the decontamination procedure kicks in. We stand just inside the door. We keep the dogs back. We take off our gloves, jackets, hats, and goggles. Then we wash our hands for a full minute—as hot as we can stand—with lots of soap. The spray is too intense, and some water bounces off my contaminated hands and splashes into my face.

Is this how I become a zombie?

I hope it was just a dry run. No virus yet.

You never really know when you're fighting an invisible army.

There are no dress rehearsals for war.

Shaken, but not stirred, I take off my two masks. The sun will disinfect them so I can wear them again. I take a shower.

Today, Shaolin is excited about her club, Salsa 5. They're going to broadcast a class with two new instructors from Venezuela.

The day passes quickly—writing and chatting with friends online. She's happy to move around and dance, and I try a little bit. I'm glad to get some exercise.

That night, Shaolin abruptly starts to scream. I rush over, but she's inconsolable. Her arm is spasming in pain, and she can't explain what happened. I realize she's gotten stuck in her sweater—her injured shoulder extended the wrong way. I help her get the sweater off and gently dry the tears from her cheeks. I massage her shoulder and arm with a balm. We're both a bit freaked out by how much pain she's in.

I worry about what would happen in a medical emergency, when we're afraid to call for help or go to the hospital.

I stay up all night listening to WHO and CDC briefings, Canadian news, American news, European news, British press, podcasts from doctors and specialists. I start listening to "city preppers" talking about how to survive "grid down" situations.

The sun begins to peek through a break in our heavy curtains, and I turn my phone off. I get a couple of hours of restless sleep and terrible dreams.

January 28, 2020 – "Life Is a Privilege"

Tuesday – Day 4

I wake up feeling tired. I have to calm the anxiety and stress, or I'm just going to make myself sick. I read some Marcus Aurelius, and this quote stands out to me: *"When you arise in the morning, think of what a privilege it is to be alive—to breathe, to think, to enjoy, to love."*

Now more than ever, we support each other. I can't change the conditions outside. I can only calm myself and do my best. *Alert, but not anxious,* I tell myself. *Relax.* I get out of bed.

It's early enough that I catch my dad in Ottawa before he heads to sleep, and we play another NHL 2020 hockey game online. I play like the devil is chasing me—and I whoop him.

Lots of friends in our Canada group are discussing safety precautions and what kind of mask can stop the virus's moisture from infecting you. I mention I'm wearing goggles too. Most people think it's too much.

Shaolin and I decide to suit up and go to the big grocery store, Ren Ren Le. On the way, I run into my hairdresser—looking dapper in a tailored robin blue suit, shiny leather shoes, no protective gear, no mask. He's got a cigarette dangling from his lips and looks at us like we're aliens.

I ask him where his mask is. Isn't he scared?

He shakes his head. *No,* he says. He's not scared. He smiles and keeps walking.

I'm never getting my haircut again.

Google says "Do the Five":

1. Wash your hands often
2. Cough into your elbow
3. Don't touch your face
4. Keep a safe distance
5. Stay home if you can

We're shopping when someone behind us starts coughing loudly. We just abandon the stalk of celery on the scale and take off.

I find some pasta sauce in what's left of the foreign food section, and then Shaolin grabs my arm. She tells me she can't see. We hold onto each other and slowly navigate through the foggy aisles.

Somehow, we make it to the self-checkout machine, bag our stuff, and pay. I'm dripping in sweat under all my layers and breathing heavily into my double masks. By the time we get outside the grocery store and into open air, I have to pull off my hood and goggles just to get a breath of fresh air. No one is around.

At home, the decontamination process takes a solid twenty minutes. I feel like I'm playing chess against billions of invisible, angry little aliens. Impossible to defend against—but if I can anticipate where they'll go, maybe I can stay sane and clean at the same time.

Shaolin's salsa class has more than 1,500 people dancing, with 10,000 watching and cheering them on. She's careful not to use her sore left shoulder, but we're happy to have something to do.

I listen to more news until late into the morning.

January 29, 2020 – "Holiday in Cambodia"

Wednesday – Day 5

My admin support worker, Nicole, tells the foreign teachers from the high school that they shouldn't return to China. Some are in Japan, South Korea, Cambodia, and Thailand. One is back home in England. They're confused—as flights are being canceled and areas are being locked down. They say they're running short on money and hoping to get back to work and paid soon—but that's not happening. Lots of confusion abounds.

The sun comes out, and we get some fresh air on top of the parking garage. No one is around, but we keep our masks on. Accepted thinking these days is that nine days is the appropriate cooldown time for any "live virus" particles, so I try to wait at least ten days between reuses, or be very mindful of how I use them.

On WeChat, people are tense and frustrated. My gamer buddy is stuck in the Hubei containment zone with his girlfriend's family. He calls me a hypochondriac—sick of my panicked messages. A good friend is mad too, saying I'm making too big a deal about what's basically the flu. It's really hard to make sure my friends here are prepared without creating panic.

There's a great quote I find online, from Michael Leavitt, former U.S. Secretary of Health and Human Services: *"Anything we do before a pandemic will seem alarmist. Everything after will seem inadequate."* I decide to post less and take care of my family more.

A journalist joins our Canadian group and starts asking questions. We ask for her help to put some pressure on the Canadian embassy to rescue our locked-down Canadian friends—except for Terry. Terry has two cats, and he refuses to leave them, no matter how bad it gets. I push him to see if Canada will evacuate his cats, but he's not optimistic.

I write 1,000 words of my book (*Amos the Amazing*). I feel blasé today. Maybe frustrated. It's hard to focus with so many people blowing up my social media.

All afternoon, I play some video games. Shaolin video chats with baby Ethan and her family, and then we watch some TV. I have the rest of my veggies for dinner.

Today's salsa class has 3,000 people dancing and 15,000 watching. It's impressive. People are excited to have something social and fun to do. Shaolin is careful to take it easy with her sore shoulder.

I cook a nice dinner for us. I eat falafels with mushrooms and garlic broccoli. I cook a steak for Shaolin. The kitchen is so smoky I start to wheeze a bit—and feel nervous about that.

January 30, 2020 – "Silent Passenger"

Thursday – Day 6

I sleep in late. It's been six days of just hanging around the house, and I feel aimless. I wash up, make coffee and breakfast for Shaolin and myself, and watch more "prepping" videos. I wonder how long we'll be in the house.

My coworkers are starting to realize this is no joke, as borders begin closing and their plane tickets back are canceled. They're making plans to return to their countries. My best friend tells me he's going to fly out in a couple of days—February 3—to the U.S., then to Canada. I kind of envy him, but I won't abandon my dogs to the streets and run away. And my wife doesn't want to leave her family. I've made roots here.

News from Germany about asymptomatic transmission backs up what China has been saying. That's scary—but it makes me feel less crazy for taking extra precautions. My friends are freaking out because they can't go outside, and all the masks are sold out.

On WeChat, Jay, an American, asks for help getting some high-end nanofiber face masks. I have a friend who makes them, and they're basically sold out everywhere. Later I realize this Jay isn't my buddy Jay—but someone else with a similar profile picture. I try to help him anyway. We discuss his plan. He's chartering a private plane to Mexico with his family. I'm a bit jealous, even though it's costing him a lot. He has two children, one just six months old. It's worth his savings to ensure their safety.

I contact a buddy of mine who still has a small batch of high-end masks in Shenzhen. He's got 200 left. By the time I get a dozen of my friends to confirm, he's down to 40. I buy 13. They're supposed to ship Monday.

It's been ten days since the market run and eight days since the family dinners—and I have no symptoms yet, so I feel pretty good about that.

We relax. The day feels like it's crawling by, but before I know it, we're having a little Chinese-style hot pot (I eat mushrooms and potatoes), and I help Shaolin cast her salsa class onto the big screen. The numbers are crazy now—growing exponentially—with thousands dancing and more than 10,000 watching and chatting.

I didn't write much today. It's hard to stay focused. I keep telling myself to stop stressing, but it's hard when I'm hiding in my home.

I sit by the window for a while and try to get some sun, but it's cloudy and grey and quiet—and we could be the only two people in the world. I've got "My Corona," like a jingle, stuck in my head. It won't go away. I download the karaoke version of *My Sharona* and record a parody cover song for my "Canadians in China" WeChat friends. They laugh. It feels good to chat with them.

January 31, 2020 – "Changes"

Friday – Day 7

It's my seventh day of being locked in at home with my wife. We try to pretend we're on a nice vacation—just with bad weather, or in a place that's really dangerous. I hope we get some good news soon.

I write my novel for a few hours, reveling in developing Amos and his amazing adventures.

Today, the Chinese specialist famous for his work with SARS in 2003—Dr. Zhong—says the virus will peak in five to fifteen days. He advises people to be vigilant and not to go outside, even if it's sunny. A few days ago, it *was* really sunny, and lots of people went out without masks, singing in the streets. The doctor told them to go home and stay healthy. The only way we can reduce infection is by staying inside.

My American coworker defies the U.S. travel advisory and flies back to China anyway—to see his girlfriend.

I get an email from CTV. They liked my print interview and want to do a live TV segment. I say okay, but it's getting late—and by the time we go to air, I'm exhausted. Still, it's nice to tell our story.

What will happen if we go back to school on February 17 and the virus is still peaking? Maybe we'll teach online for the month. We really don't know anything, and our school has no idea either.

We just have to wait and see what happens.

February 1, 2020 – "No Saturday Night Fever"

Saturday – Day 8

It's Saturday!

I do some pushups and weights, and drink Italian coffee from my French press. I stretch my muscles. I hope I can get back to the gym soon.

There are now more than 11,000 cases of coronavirus and pneumonia in China, and I can't determine if this means that every case of the virus also develops pneumonia—which would make the situation much more severe. It seems highly improbable, but a translated article suggests this, so we seek clarification. In Chongqing, we have 238 confirmed cases. One local person has died. One has recovered and been released. I hope we see more of that soon.

I pull out my ukulele and play for a while, but get distracted—fiddling around and doing some work on my laptop: writing, editing. I wish I were doing this from the beach. What a dream.

I watch some YouTube livestreams of my favorite beaches and feel happy—and then a bit jealous.

I binge on podcasts and YouTube videos, and we have hot pot again.

We do a salsa class online with Shaolin's phone, and it's fun.

February 2, 2020 – "Everything Is Closed"

Sunday – Day 9

I talk to my dad. He's worried about what's going on, and it makes it hard to talk with him. I'm impatient. I can feel it in my bones—it's going to be a hard day.

I get suited up for a walk. My wife is craving Oreos, and I want some bread. I decide to shoot a video of my trip so my friends can see. I feel like an astronaut.

Everything is closed on my street, and the only bakery that was open closes just as I walk by.

My good friend Sean moved up his plans to go to Canada because the U.S. is canceling flights from China. He's already in Los Angeles and making his way back to Toronto.

My wife and I are feeling a bit stir-crazy, so I go to work in my office. Then I get into a flow state and write close to 10,000 words. That's really good. She comes in—and I hate being disturbed while I'm writing—so we get in a bad argument. She smashes a glass bottle on the ground. It's hard to be cooped up for so long. We need some fresh air and sunlight soon.

Later, I talk to my good friend Stu about healthy marriages. He's full of positivity and sound advice. Being a productive adult seems like one must be willing to make sacrifices. Being a good husband seems like you just keep going when you're having a tough day.

February 3, 2020 – "Every Meal Is a Decision"

Monday – Day 10

I get out of bed and wash my hands, wash my hands, wash my hands—*myyyyy haaaaaaaaaands*. They feel like raisins. They're prunes. I've never been so clean.

I'm sick of *Happy Birthday*, so I sing *Alouette, gentille alouette*, but I can't remember all the words, and Shaolin makes fun of my French accent. I transition to "Bohemian Rhapsody" because it's like ten minutes long—and if Queen can't get my hands clean, then I guess I'm just a dirty boy.

It feels like I wash them a hundred times a day—before and after every task—so I can have the pleasure of scratching my nose if I choose and not worry it will make me sick or kill me.

Yesterday, the bakery closed down, so I'm not sure if I'll be able to make toast anymore. I try to order some flour and cheese to make pizza, but my foreign passport doesn't work for the Chinese app system's ID requirement, and Shaolin doesn't want any.

The balance between wanting food to last and not letting it go bad means everything is now a serious decision. I'm storing boiled water in containers in case I can't order more spring water—and in case we lose power or water for some time. I'm pretty sure that won't happen. But being prepared is the only antidote to feeling helpless.

Shaolin is still grumpy. She's watching a video about a young Chinese woman who goes to Santorini to take wedding photos—but doesn't have a boyfriend. Her driver is laughing at her.

Shaolin and I took wedding photos from the top of the mountain in Oia Village, Santorini, last summer. It was an unforgettable day, but I framed the photo anyway—for when we're too old to remember.

We get a call. A package with some toothpaste is waiting for me downstairs. Shaolin—always skeptical of my apocalypse-hoarding instincts—does order a little more than we need to placate my inner spirit squirrel.

I suit up and head to the only remaining open market. My goggles are so foggy by the time I get there I can hardly see, but I find a loaf of sweet bread. I let everyone else go first so I don't have to crowd next to anyone, then pay with WeChat's cashless system.

Outside, I climb the overpass when the sound of hacking makes me stop and look down. Twenty meters away, near the market I just left, an old man is kneeling. He's wheezing, hacking, and spitting on the ground. It's shocking. I freeze.

An older lady walks out, and they walk away together.

No salsa class today. Shaolin doesn't feel like it. Same for dinner. Maybe we'll try again tomorrow. I missed dinner too. I feel a little hungry, but not enough to make anything.

Finally, around 10 p.m., I cut some slices of bread and make pizza toast with pasta sauce, feta cheese, and fried garlic. It's not bad.

I think things are going to be okay.

February 4, 2020 – "Rationing and Preparing"

Tuesday – Day 11

I'm running low on coffee. I think I'll go without it today.

The Stoic philosophers of ancient Greece and Rome practiced "going without" to enhance gratitude and ease anxiety. If I can be okay without coffee today, I'll be doubly grateful for coffee tomorrow.

I binge on MasterClasses, trying to make the most of the week before my account runs out. Today it's a writing and comedy class with David Sedaris. He's hilarious—really witty. I learn the importance of asking smart questions and turning small talk into something meaningful. A precious gift.

The newest numbers for Chongqing show 337 cases of acute coronavirus with pneumonia. That's what we're calling it now, even though it's puzzling. I research more about RNA viruses to better understand what we're up against.

My district is now the second highest, with 15 confirmed cases and one death. A student's mother told me it was a restaurant worker near my campus. It's scary. We're grateful to be cooking at home.

On social media, posts of compounds with infected residents are making the rounds. One of my good friends and his wife panic when they hear their compound has a dozen infected—until they realize it's a building with the same name in Wuhan.

New measures have been announced to protect the people of Chongqing. Anyone returning from outside the city, from other provinces, or from abroad must self-isolate or be quarantined and observed for 14 days. The same goes if there's already a confirmed case in your residential area, or if you're connected to a previously confirmed case. It seems strict, but no one's complaining.

In our Canadians in China chat, someone asks whether microwaving a cheeseburger made by an infected person would kill the virus. Others tell them to relax and enjoy their lunch. Still, one restaurant worker in town was confirmed infected and spread it to all his coworkers. We don't yet know about the customers.

We're all worried—but we're hopeful. The actions China is taking are unprecedented, and they'll have a huge impact in breaking the chain of transmission—*if* people can be patient and stay home.

Official numbers today show:

- 20,471 confirmed cases
- 23,214 suspected
- 2,788 severe
- 426 deaths
- 657 recovered and released

It's a big jump from just a day or two ago. I read that without social distancing, the numbers could be **68 times worse**—millions of infections averted thanks to these strong actions across China.

Mid-afternoon, I cave and make a pot of coffee. It's so savory, so aromatic and delicious. My headache fades. Tomorrow, maybe I'll try again to go without—maybe every second day.

Thirteen fancy nanofiber masks arrive from Shenzhen. We got a call from Wuhan and briefly panicked—*if they're coming from Wuhan, never mind!*—but only the admin worker was calling from there. The masks were shipped from Shenzhen. I send two to Chengdu for my British friend and his pregnant girlfriend.

There are more stories to edit. Community leaders are going door to door, knocking and checking in on people in their grid to make sure symptomatic patients are found and treated. I hope they're taking proper precautions.

People feel increasingly optimistic that we can isolate the infected and beat this.

My spirits are high today.
Zhongguo, jiayou!
Go, China!

February 5, 2020 – "Happy Birthday, Ethan!"

Wednesday – Day 12

Happy birthday to you! Shaolin sings to baby Ethan on video chat, and he claps his hands. He's been loving the song ever since he turned one.

I wash my hands with soap and water and count two *Happy Birthdays*—twenty seconds. That's what the government recommends to make sure we kill the virus.

Today, China reports 24,423 cases of infection—3,949 more than yesterday. In my city of Chongqing, we now have 376 confirmed cases and two deaths.

Our good friend Chris gets a call from the police today. She was shopping at a supermarket on January 23. At the same time, a person from Wuhan was shopping there and was later confirmed infected. Chris reports no symptoms, but they still ask her to stay home for a couple of weeks. It feels like it's coming closer, but inside our house, we feel safe.

As Chongqing braces for a rise in infections, families are being asked to allow only one person to leave the house every two days—to buy vegetables and then return.

I go shopping again today. At the door of the supermarket, a worker is baking her mask in front of a space heater. She shoots me in the forehead with a temperature gun but can't get a reading. She asks me to take off my goggles. I politely decline. She scans my forehead, then my ear, then the side of my face. Finally, she asks me to remove a glove. I offer my hand, and she gets a reading: 36.4°C.

"No problem," we say to each other, and I carry on.

I peek at the stall where the live chickens used to be, but they're gone. Only fish on ice remain at the back.

On the way back into the school compound, the guard shoots me with a temperature gun. I'm worried because I'm hot and sweating buckets from hauling heavy bags—but they let me in.

My school sends the foreign teachers a check-in questionnaire about our health. Local community leaders are going door to door, reporting symptoms and arranging medical treatment. Most people are happy to wear masks in public, limit their time outside, and seek help when needed.

Some, however, aren't so cooperative. They scream and fuss in indignation until the police arrive. That's usually when their fury softens and the excuses begin—right up until they're dragged into a police car and driven away. These videos are shared publicly as a form of no-mask shaming—to show that disobedience won't be tolerated at the risk of public health.

When the house is on fire, it's not the time to discuss politics.

Lead, follow, or get out of the way.

I read. We watch TV.

Andrea drops by to pick up his face masks. We stay two meters apart—though we bump elbows when we say goodbye. He took the bus over, which shocks Shaolin. He said he was alone, and the bus smelled like disinfectant.

The bus!

I'm starting to teach online on Monday, but the system is complicated, and I don't understand it yet.

I hope I can still come to Canada this summer. My grandma turned ninety this year, and it's essential I see her. I give her a call, and she's happy to hear from us.

I made it a whole day without coffee. No headaches.

I can't wait for tomorrow.

February 6, 2020 – "Debunking a Rumor"

Thursday – Day 13

I stay in bed as long as I can, savoring the memory of normalcy.

Current infection numbers in China are close to 28,000 today—376 of those in Chongqing. A big jump nationally, a smaller one locally. But the infection is close to our home now. We have an app on our phones that shows where the local cases are. The closest is just half a kilometer away—two infected, at the shopping mall. It looks close. There are seventeen cases within five kilometers.

Shaolin asks me not to leave the house anymore. We have enough food to last a while.

My coffee is delicious.

The newest scary rumor hits during a video call with family and friends: a restaurant worker near our home dropped dead—no symptoms. Later, I debunk it. Turns out, the 61-year-old woman had hidden her symptoms for eight days before collapsing. She was admitted to the hospital and later passed away. Still a tragedy—but easier to process. Her coworkers and customers are being tracked.

Today's MasterClass is Neil deGrasse Tyson, one of my favorite science educators, teaching about the scientific method. It's medicine for the anxiety and wild rumors going around. It's excellent.

A local community wakes up to the sound of snoring over the neighborhood broadcasting system. My friend sends me her video—it goes viral. A good laugh is healthy.

Andrew, a Canadian in China, is making sour cream and perogies from scratch. When "What's for dinner?" becomes the big question of the day, recipes are hot gossip.

Canada's embassy now recommends all Canadians leave China. Not helpful when flying seems more dangerous than staying home. I'm not going to abandon my pets and run.

The big news in treatment: Remdesivir and Chloroquine seem promising for treating the novel coronavirus. Hopefully available by April. Fingers crossed.

One friend left for Vancouver via Hong Kong today. Another went out for groceries and came back to find his building sealed. He's in a hotel now.

Between my diary and journalism work, I'm busy. I start prepping to do video classes. Our new ETA for real classes is March 1.

We tidy the house. Shaolin tries to put the mini vac back together and pulls her shoulder. I find her crying in the bathroom and take care of her.

An hour later, she's back in bed with ibuprofen in her belly and a hot pad on her arm.

Gotta be careful—we're on our own.

For dinner, we share extra-crispy homemade fries, lentil soup, and steak for Shaolin. I exercise while we watch movies, and we go to bed.

February 7, 2020 – "Stocking Up on Food"

Friday – Day 14

We are full of uncertainty. If we hadn't gone outside for supplies, we'd be confident we were healthy. Panic and rumors make us nervous. On the news, a man who removed his mask for fifteen seconds with a cashier at a supermarket is now infected.

I never took off my mask inside a store. Still, I can't be sure my de-contamination protocols for clothes and incoming supplies have been enough. Every cough, sniffle, and sneeze arouses concern.

Shaolin and I debate stocking up now, before Chongqing peaks. I let her win the argument—but lo and behold, she orders three big bags of groceries anyway.

That's love.

We're going to bar the door and wait fourteen more days.

We're trying to make a plan for the food. I have a bottle of red wine I'll save for Valentine's Day next week and will make a nice dinner. We stay busy all day. If you strip away the fear, anxiety, and panic—it's like hitting the jackpot for an introvert.

Last night, while falling asleep, I was overcome with regret for leaving two unpublished manuscripts in my drawer for so long. I worried I'd never have the chance to publish them.

It made me think of Ernest Hemingway's incredible short story *The Snows of Kilimanjaro*—where he writes about a writer who dies of a leg infection near Kilimanjaro, lamenting all the stories he never took the time to write. That he put off until later. And now, dying, realizes he will never give them to the world.

There is a creeping horror about dying unfulfilled that is much scarier than just dying. The death of a dream, of hope—that's worse. I don't fear death, but I fear an unfulfilled one, I realize.

I get out of bed and outline my own "Snows"—about the struggles of maintenance and unfinished dreams. I hope I'll find the time to write more, if I can calm my mind. I've always felt I had a good book in me—if the world would just leave me alone long enough to write it. Maybe this is my chance.

If I can stay healthy long enough.

That's how a writer's brain works, I guess.

Today's a busy day working on local news. Lots of coverage about procedures for getting people back to work, and a protocol handbook on public health. I fight hard to amend one confusing passage about coronavirus transmission and pets—not specifically about *novel* coronavirus (nCoV).

There are already tragic reports of people killing or abandoning their pets in panic. Both the World Health Organization and local experts say there is no evidence of infected pets carrying nCoV. We amend the section, and our station chief passes my concerns up the chain to the health board to clarify the Chinese protocol guidebook.

I hope, in some small way, it helps ease the suffering of our furry friends. These battles have left me exhausted, my empathy batteries drained.

Shaolin makes little paper finger puppets and entertains baby Ethan on video chat. We play and laugh with him. I eat lentil soup and potato salad, along with a couple of delicious German beers.

I find myself thinking about Dad's plump, red, fresh tomatoes and the fragrant perfume of lavender in his Ottawa garden in the summertime. I pick up and sanitize our delivered groceries and take a shower.

Later, I listen to the news. Quarantined cruise ships and the Wuhan plane touching down in Trenton are dominating the cycle. I can feel patience waning. The international media seems hungry for something exciting to report.

I spend a few hours with Margaret Atwood's creative writing class. Her voice is one of a kind. No matter how this goes, I promise myself: I will post and publish as much of my writing as I can this year.

Nothing like the fear of death to get you in gear.

February 8, 2020 – "A Motley Crew of Astronauts"

Saturday – Day 15

We are all astronauts—a motley crew—traveling through space at 268 kilometers per second on a planet-ship equipped with gravity, a vast array of dining options, and Broadway entertainment.

Our solar system spirals like a DNA strand, like a Cavatappi noodle, rocketing through the infinite darkness—from Lyra to Betelgeuse to the Magellanic Clouds to Andromeda and beyond.

Unfortunately, we don't work that well together. And if we keep scrapping our life support systems for the promise of imaginary "money," we're not gonna make it. We are clever monkeys. But we are not wise.

The cruise ship situation off the coast of Japan looks scary. Shaolin worries about recycled air coming up through pipes and infecting people in apartment buildings. The social media comments, if not the science, regurgitate that fear.

Chris Hadfield teaches me that astronauts don't cross their fingers—they manage risk. Being afraid is a choice. A virus isn't scary. But many people are scared. Why?

The unknown.

Can we learn, take precautions, and manage the risk—choosing not to be afraid? Of course we can. So today, thanks to Chris Hadfield, I became an *AsCan*, or honorary Astronaut Candidate. AsCans train for anything that could go wrong, because out there, you're on your own. You have to be able to solve the problem—or get to safety—in one breath.

This works for me.

My dad calls, and we play a game of online ice hockey—something we've tried to do a lot over the years since I moved to China. Just a few minutes, but it feels good to stay connected to my homeland, and my family.

I squeak out an overtime win.

I dig up an old reference letter that got me to China. That helps my former boss Judy draft a new reference letter for me. She recruited me six years ago but left our school recently. And now seems like a good time to have all my ducks consecutive.

The situation is fluid.

We tidy the kitchen. The red peppers, some carrots, two tomatoes, and an apple are moldy. It pains me.

Today is the Lantern Festival—the final day of Chinese New Year.

The magic word is: *pangolin?*

We call the family. Baba is happy—he's played Mahjong every day for the past two weeks. Baby Ethan is doing great too. After all, it's winter. Shaolin makes hot pot for dinner. Around midnight, we eat glutinous rice balls filled with black sesame sugar.

A friend who's been going out and dining in public blows up a group chat, ready to panic. While eating with someone, he "learns" that nCoV has become airborne.

After spending a good hundred hours researching this coronavirus, I shut that down. There's no evidence of airborne transmission. It's likely a miscommunication caused by poor translation.

We're good.

For now.

February 9, 2020 – "Bad News and Worse"

Sunday – Day 16

I'm quite bored.

I don't feel like writing anything.

We do our morning rituals. I decide to drink instant coffee today and save my beans for a sweet treat once a week—until I can order more. I can't bear the thought of running out.

For brunch, I sprinkle chia seeds on my salt-and-peppered boiled eggs and smashed avocado toast with fresh spicy hummus. I've lost three kilograms while getting minimal exercise. I guess giving up fast food makes a difference.

My editor tells me that an information official in Chongqing wants to adapt my blog into Chinese as an excellent example of how to stay safe and take suitable citizen precautions to reduce viral transmission.

I feel like a volunteer in (viral) wartime.

It is an invisible war.

I'm a special agent in the field of "sitting on my ass."

I can train others to sit on their asses, too.

The Canadians in WeChat are discussing two pieces of news released by Chinese media today. The worst news is that a girl apparently returned from Wuhan and didn't develop any sign of the infection for twenty days. If true, this could have global implications. Don't tell the stir-crazy Canadians on that cruise ship.

There's worse news.

Aerosol.

The media reports that aerosol transmission is possible. That means if you walk into a sneeze cloud or a public bathroom where someone is using the toilet, you could become infected.

It's terrible—but I might be the first person to put these pieces together: the 2019-nCoV virus can live in the intestines and fecal matter... and it can be aerosolized.

What does that mean?

It means we have to watch out for virus farts that can give you pneumonia and then kill you. This shouldn't be funny—but I laugh really hard anyway, feeling like a crazy person. People stop outside my window, and I whisper an apology for ruining the solemnity of their moment with my cabin fevered dreams.

Silent but deadly becomes a reality.

2020.

My mom calls me crying because she read my diary, and it made her feel sad and worried. I cheer her up with horrible jokes, and she laughs until we're okay again.

I burn my hand making toasted garlic broccoli and pickle pizza sandwich burgers. It's not bad, but it might leave a scar.

Scars make a man, though.

My life is like an elaborate performance art of Kintsugi—◈◈ (*jīn jì*)—broken, yet glued together with gold, and somehow all the better for it.

I play some video games, watch some news, and listen to my daily medical blogs. When I'm overcome by ennui, I tune my tenor ukulele and play until my fingers bleed. I cut my nails when I realize how long they've gotten.

I'm getting pretty good at *Danny Boy*.

We eat nacho cheese chips with salsa and hummus dip while watching *Jojo Rabbit*. Tomorrow, my bean sprouts will be ready to harvest.

Life finds a way.

February 10, 2020 – "Rock Show for Babies"

Monday – Day 17

We wake to abrasive screaming through the broadcast system, muffled through our closed windows. A voice tells us not to go outside because the virus is not finished.

It's a sunny day, and people are restless.

Dr. Zhong posts a message: our first quarantine failed because too many stubborn people kept socializing. He warns that we'll need a better one to truly stop the spread. Around the world, more cities go into total lockdown today. The drastic measures scare some, but most see them as necessary to combat this virus on a global scale.

Today is the first day back to work for many people in China. They're trying to get life going again, getting factories up and running.

One of the last great Roman Emperors, Marcus Aurelius, wrote a helpful exercise. You should view your life from above—like a bird's-eye view of your home, your community, your city, your country. As if the whole planet is just a blip in the endless vastness of space. Your base emotions—anxiety, panic, fear, anger, frustration, boredom—are not so important. We are but small players in a much larger story. When seen from above, our feelings shrink and disappear.

Together, we fight to extinguish the chains of viral transmission. *Multum in Parvo.*

The micro stops the macro.

A virus many times smaller than this . stops the world.

More Canadians are trying to leave China. Thailand is one of the few places still offering (cheap!) flights. I keep seeing people running from containment—only to carry the disease with them. In centuries past, ships were required to remain offshore for forty days to prevent outbreaks. *Quarantine* comes from the Italian *quaranta*—forty.

But the humans would wait on the ships. The rats on board would jump into the water and infect the new land.

People can be so selfish when it comes to their own convenience.

We are all safer if we stay put.

In a comforting reversal from yesterday's reports, experts now say there's no evidence for aerosol transmission. Still, cities continue to spray public areas with strong chemical disinfectants.

If my day feels full of paradoxes—it might be.

We can only wait.

High school starts today, online. My first English class is an audio call filled with glitches, echoes, delays, and giggling voices. Some students are busy with homework and hardly noticed what's going on outside. Others are angry.

"Angry at what?" I ask.

"Those people who ate bats and caused this."

I let them know we're not sure bat soup is the cause, but they've heard what they've heard—and they're mad.

The delicious bouquet of my lentil soup masks the spicy kick. We eat well today.

I'm practicing my ukulele and Shaolin turns the camera on me. Next thing I know, we've got a dozen family members watching our rock show for babies. Shaolin is an ethereal singer. I strum through half a dozen children's classics before my fingers swell up.

February 11, 2020 – "Go Inside. It's Not Safe"

Tuesday – Day 18

It's dark. I'm feeling woozy and I don't know where I am.

Shaolin is telling me to get up and close all the windows—trucks will be spraying the city with disinfectant chemicals. I run around the house, slamming windows shut. Before I pass out, I send a few messages to my friend groups on WeChat.

Around 10 a.m., I get up and check the news from inside the warmth of my pillow fort.

They ask, *"Did it happen?"*

I ask Shaolin if she's seen any pictures from her 5,000 friends on WeChat. She has—videos—but she's already deleted them.

My friend is angry, telling me to stop spreading fear. I apologize. I stop posting in local groups.

A lot of chat groups—our only form of socialization for weeks—are fracturing.

Tensions are high.

Another broadcast warns that the virus can travel through water pipes, as SARS did in Hong Kong in 2003. People are covering sinks and drains. I'm caught between hard science trying to catch up and well-intentioned ideas that might not be accurate.

Dr. Zhong, the Chinese epidemiologist who discovered SARS in 2003, releases a troubling study. It says the incubation period for 2019-nCoV can last as long as **24 days**, and half the patients tested didn't have a fever when they sought treatment. This has serious implications for testing standards in many countries.

Big bummer.

We're sitting around, bored, when a mischievous sunbeam drops by. We jump up in excitement. Outside, children chatter and basketballs bounce against concrete. A booming voice on the loudspeaker yells in Chinese:

"Go inside! It's not safe! Return to your homes!"

Soon, it gets quieter.

We grab two chairs and sit on the rooftop of the upper parking garage. We're all alone, but below, I can see other little clusters of sun bunnies—stretching, basking in their own private squares of light.

Shaolin salsa dances in the sun, and I play *Shadowrun* on my phone with Andrea. A couple walks by with a little grey cat, but they keep a respectful distance. On the street below, a man smokes a cigarette and hands off a package of masks to a grateful woman through the gate.

After a couple of hours, we head home.

I shower and put on clean clothes.

I enjoy spending my days writing, and I'm not sure if I'm ready to return to the hectic classroom. Will it feel safe? If factories can reopen today, why are we still looking at March for school? Will I be standing there with goggles and a mask, teaching a bunch of students in bubbles?

February 12, 2020 – "The Negotiator"

Wednesday – Day 19

A lot of friends send messages to me about staying healthy.

What they don't understand is that this solitary life is peaceful.

I haven't had a sniffle for a month, but this morning Shaolin is coughing. She has a headache and naps through half the afternoon. I drink strong coffee and teach another class online.

The virus has an official name now, thanks to the WHO: COVID-19—designed to be forgettable.

Shaolin learns how to make pancakes. I make a "Quarantinos" pickle pizza.

I outline, I write, and when that mythic, sweltering Chongqing sun lights me up, I sit on the balcony reading Stephen King until my face starts to burn.

Later, I take former FBI negotiator Chris Voss's MasterClass. He gives me an idea.

This virus has taken China hostage.

We need a negotiator.

"Hey virus," I begin. "I'm going to call you COVID-19. You can call me Kai. I'm here to find out what you need."

Gnashing teeth. Blazing red eyes.

I'm a ferocious, virulent coronavirus.

I use the late-night DJ voice. "It seems like being taken seriously is valuable to you."

I'm a world-class pandemic, it roars, puffing out its chest. *I'm gonna dwarf the Spanish Flu.*

"The Spanish Flu?" I mirror, voice falling downward.

A century ago, they didn't even have planes. Today, I've hopped around the world to 24 countries. In a week I'll be in 100 more.

I don't ask *why*. I ask *what*. Asking why will make COVID defensive.

"What about traveling the world is important to you? It sounds like you want to be a big deal. Be remembered."

That's right.

Better sit up. Pay attention, it snarls, jagged crowns for teeth.

I need to find a black swan.

"What most worries you?"

That's it. I see the fear in its eyes.

"Ah, I see. We're all holed up in our houses and we're bored, but you can't spread the way you want to."

COVID shifts on its paws.

"What are we trying to accomplish?"

You quiver in fear, it says. *And I'm outside, waiting for you.*

Tactical empathy.

"How am I supposed to carry on with my life like this?"

Go outside. I dare you.

Drop an anchor, Chris whispers in my ear.

"So let's say I go outside. What are we looking at? Fifteen percent complication rate? Two percent fatality? The odds are pretty good for me."

COVID growls and chomps its fangs.

Then it hits me.

COVID is a sensitive young Libra—terrified of being boring.

The virus is a paradox: cool, clinical, sterile... yet secretly dramatic. It craves attention.

"I want you to feel like you're being treated fairly. We respect you. You've ground the world's factory to a halt. We will never forget you."

Bend their reality. Anchor them in preparedness for a loss.

"But let's be honest here—you're nasty, but you're not the zombie plague. We *can* get over this. Keep this up and that universal vaccine gets to market faster. You're bigger and worse than SARS. I'll give you that. COVID, we're going to remember you."

COVID nods—feeling respected.

Appreciated.

"How about we use your memory to help us prepare for the next big one? You'll be our textbook example of why we must remain vigilant. When we wash our hands, when we refuse to scratch our itchy noses, your name will be on our lips."

...And yes, you can kill a lot of people.

It stops gnashing. Considers.

Use leverage.

"I can make you famous. But you've gotta calm down—and let us get our factories up and running again."

Ninety minutes later, we have a deal.

February 13, 2020 – "A Trip to School"

Thursday – Day 20

Shaolin is feeling much better. We've been getting good sleep.

Doctors say the key to fighting and beating this kind of virus is a strong immune system, and 7–9 hours of sleep a night can mean the difference between resting at home and fighting for your life in the ICU.

I feel for the thousands of exhausted healthcare workers in Hubei—many of whom have already become infected—and I'm proud that Chongqing, along with so many other cities, can send support and resources to the epicenter.

I'm aware that if new clusters break out across China and we lose the ability to focus our efforts on one hotspot, it would be a very scary situation. So I'm determined to do my part and stay home.

Many people are back to work here in Chongqing, with an attitude of optimism. Many of us are still on lockdown too.

We tidy up, air out the blankets in the sun, and eat well today.

A change in testing criteria—using CT scans and other diagnostics—means 15,000 new infections are on the board today. It feels like a lot. Now there are 218 infected on the *Diamond Princess* cruise ship in Japan. They must be so stressed. At least those of us quarantining at home have our comforts.

My coworker, an American teacher holed up across the country, needs help getting lesson plans off his laptop. Our school office has a set of keys to his flat, and he talks me into doing a little break-and-enter (with permission).

Shaolin wants me to stay home, but I get suited up and walk over to the school gate.

The guard comes out and waves me away. I call Shaolin to help get more information, but before I can back away, the guard grabs my phone.

The school has been decontaminated. No one goes in. That's it. Rules are rules.

I return home, remove my phone case and leave it by the door, spray everything down, shower, and change my clothes.

Today, we see a lot of videos of police arresting citizens who create panic and spread rumors. During any crisis, some people will take advantage of the chaos for their own profit.

I want to keep my head down, stay quiet, and just make it through this. I'm here to be a teacher, and I'm employed as a writer.

Like Popeye said: *I am what I am.*

We play another WeChat concert for babies.

I start reading William Gibson's *Agency* and Laurie Garrett's *The Coming Plague.* They're both fascinating for different reasons, but it strikes me that I'm juggling a lot of balls.

Maintenance is always the hardest part for me—but it's where I'm working the hardest to develop.

A few more friends leave for Vancouver. They're lucky to get on one of the remaining flights out of Shanghai. I hope a trip to Canada this summer is possible.

The World Health Organization and many countries brace for more spread.

February 14, 2020 – "Lovers in Dangerous Times"

Friday – Day 21

I make Shaolin breakfast in bed: pancakes and coffee. Her shoulder is starting to feel better.

It's a sunny day, and we tidy up, take care of our plants on the balcony, and move our trees around the living room so they can soak in more sun.

I go outside three times today, which is unprecedented.

The first time, we take a couple of chairs to the parking garage roof and bask in the sunshine. We see a few people leaving the school with some kind of paper pass. We don't have one—no one seems to realize we're the only occupants of the foreign teachers' dormitory building on campus. It's fine. We don't want to go anywhere.

After a couple of hours of fresh air, we pick up Shaolin's shoulder medication from a tiny automatic mailbox on campus.

The garbage outside my house is piling up and stinks to high hell. It seems the workers, like everyone else, are sheltering in place. But people are still tossing their trash into the large bin near our front door. Cooking in the sweltering sun, it's basically an invitation for rats and disease. I've chased rats away from my front door before, but it'll be harder now.

We decide to order food made by unvetted strangers and delivered to our house. It feels dangerous—and romantic—in a star-crossed lovers kind of way.

Shaolin wants to order KFC. I'm a vegetarian... more by preference than rule. I've eaten giant spiders and scorpions in Asia before. Today I'm feeling sentimental.

My dad used to call KFC "Champs Chicken" when I was a boy in Ottawa.

When I was 20, we flew to Winnipeg for Grandpa Wood's funeral. I saw his garage full of curling trophies and newspaper articles. Howard Wood Jr. and his father, Howard "Pappy" Wood Sr., were legendary sportsmen—inducted into the Canadian Sports Hall of Fame, the Curling Hall of Fame, and even the Guinness Book of World Records for most bonspiels.

I remember a photo in the *Winnipeg Tribune* of them winning the Brier in 1940. There was another one from 1947, the four-man team looking so proud, standing in front of four gleaming new Hudsons.

Everyone says I look a lot like my grandpa.

So, I pick up the KFC at the gate. Then I go out again—third time—because Shaolin's Baba has sent us another package while we're preparing for our next online class. He raised a duck in his rooftop garden, baked it, and shipped it across town as a gift.

The new video class is just getting started. I've got my phone and laptop set up, and I sit down on the couch.

It takes a minute to realize the sofa is wet.

Without being weird, as six ten-year-olds and their families watch me introduce myself for the first time, I realize I'm sitting in dog urine. Ben Ben, my old poodle, is angry I went out three times on a sunny day and doesn't understand why I'm not walking him anymore.

I wish I could tell him it's for his own safety.

So, I sit in a puddle of dog pee, practicing stoicism while teaching screaming children without complaint.

Hey—it could be worse.

We finish class with a song.

After class, I still give Ben Ben a good rub. The boy is getting old.

We watch *The Terminal* with Tom Hanks and laugh at how outlandish his version of quarantine is.

PART II MOVEMENT

February 15, 2020 – "Hoping for a Vaccine"

Saturday – Day 22

I wake up and drink instant coffee. Unsatisfied, I make a big pot of freshly ground beans to savor by the sunny window. We've put in an order for more with an unknown ETA, but with things returning to "normal" out there, it may actually arrive this month.

Today we're doing six full hours of online teaching for Shaolin's private students, making Saturday my biggest workday of the week. The first class runs from 10 a.m. to noon—they're a sweet bunch. The technology is iffy at times, but we make it work with a few breaks and a tasty ukulele session at the end.

I make some guacamole on toast with boiled eggs for brunch.

Tomorrow, I'm going to have to make my own bread for the first time. Exciting and strange.

There's something eerie about being the only ones in an empty building—forgotten and alone. My neighbors, the foreign teachers of Chongqing Foreign Language School, are on perpetual "shelter in place" orders in vacation countries or back home with their families.

Still, at a time when contact is dangerous, there is safety in solitude. I'm trying hard not to panic, so I'll limit my doom and gloom to one stream of consciousness.

In good news: Chinese hospitals are using blood plasma from recovered patients to try to create an antibody response and boost immunity. Vaccine tests are proceeding well, and antiviral drugs like Chloroquine and Remdesivir seem to indicate a quicker recovery from the virus.

In weird news: It's possible that monkeys re-exposed to the virus may trigger a *cytokine storm*—an overproduction of immune cells and their activating compounds. In this scenario, the body's own defense system starts attacking its own organs. Some experts fear a secondary infection could be more destructive than the first.

We saw this a century ago with the Spanish Flu—which, by the way, didn't even start in Spain.

I hope a vaccine comes along before summer. I'm getting tired of feeling like I'm trapped in a horror movie. When the stress creeps in—like painful knots through my back and shoulders—I hold my dog Ben Ben and we both breathe for a while until everything feels like it's going to be okay.

On the plus side, the virus epidemic has temporarily reduced my anxiety about climate catastrophe.

The second class today is another new one. It goes pretty smoothly.

I read *Agency* by William Gibson during a two-hour break—it keeps me entertained.

Our third class is good, too.

Today is exhausting, but I guess that's why they call it work.

February 16, 2020 – "Space Walk"

Sunday – Day 23

I might be in a simulation—or on the peripheral of some crazy-rich, historical-crisis tourist's immersive tour from the future. I can imagine this "extreme disaster package" would be quite expensive.

In feel-good news: a man from Wuhan is volunteering with an animal rescue group to care for nearly 5,000 pets whose owners are locked out of the quarantine zone. Some give passcodes; others beg him to break in. He fills food and water bowls and changes cat litter. A quiet hero.

The school has opened the admin office for a few hours today to help residents avoid a blackout. In China, we prepay for electricity. There's no grace period. You either have kWh—or you don't.

In this absolute system, we need them to open the office to recharge our power cards.

I suit up for the outside world: mask, goggles, gloves, and the outdoor clothes I keep by the door in a "hot zone."

At the security gate near the middle gate, I drop off my power card and some cash. Stupidly, I didn't bring my own pen and sign the form with theirs—then walk away, wondering how many other people have used it. I'll have to come back at 4:00 to pick up my card "contact-free."

I decide to take a walk to the street.

The turnstiles are closed, and I don't have a pass to go out, but I manage to wiggle through the vehicle gates and walk toward an SF Express delivery truck (Shunfeng 顺丰, like FedEx) parked out front. A young man is sorting packages. I snap a few photos just as two community leaders walk toward me. They gawk at this hulking alien in a bug-eyed spacesuit taking pictures of them.

I scuttle back inside the compound before they start asking questions.

I decontaminate, wash up, and get my "inside onesie" back on.

Dad is still up, and we play a game of NHL Hockey online. I'm Team Canada, and he's the "All-Time All-Stars." Crosby versus Gretzky—a timeless matchup of ice giants. When I was in college in Halifax, my best friend used to do Sidney Crosby's manicures in Cole Harbour. The guy really cared about good hand hygiene. That was the year *The Root Sellers* played the 2010 Vancouver Olympics (Yukon medal ceremonies), where Sid scored the Golden Goal against Team USA.

I take my morning vitamins and finish the last of my "zinc immunity booster" pills.

The gas runs out for our stove mid-pancake, but I finish cooking on our little hot plate.

Shaolin shares a story about a man on Day 13 of his 14-day quarantine who goes for a walk in the park. Suddenly, his boss calls him: *Go home now.* Facial recognition flagged him. An AI system notified his employer and pressured them to send him back.

China's use of AI and big data to combat the spread is terrifyingly efficient. I wonder how countries without that level of tech will manage something like this.

Millions of roses were destroyed this Valentine's Day.

The top three most popular gifts this year?

Face masks, eye goggles, and alcohol-lathered cotton pads.

Love in the age of no contact.

Love in the time of coronavirus.

February 17, 2020 – "Shooting Star"

Monday – Day 24

This is all my fault.

I'm sorry. I didn't mean it.

I wished on a shooting star—for this Spring Festival to be relaxing, long, and productive. I wanted a break from the endless sore teacher's throat and to spend more of my energy writing prolifically.

Now look at this mess.

A friend suggests I read a 40-year-old thriller by Dean Koontz called *The Eyes of Darkness*. In it, a novel virus is released in Wuhan, China. Eerie. But I suppose if you search hard enough, you can find prescience anywhere.

I mean, *The Simpsons* predicted Donald Trump as president.

I smash up some guacamole and go from 0 to grilled cheese in 60 seconds. My fresh-baked bread blows my mind. Why have I never done this before?

Oh yeah—I had access to a bakery.

School hours are now further reduced to limit stress and eye strain. After lunch, we have an online class, and my students are happy and relaxed. Some of them made beautiful, creative short films about their "special time staycations."

I'm happy today—and thinking of a Stoic quote by Epictetus:

It is quite impossible to unite happiness with a yearning for what we don't have. Happiness has all that it wants.

No regrets today.

As far as Mondays go, this one is pretty painless.

I still have concerns about people going back to work, but the gas man shows up at the school gate to swap in a new canister. He gives me a paper receipt, and I hold it awkwardly in a gloved hand. Not wanting to bring anything "foreign" inside, I pass a tree and tuck the receipt into a hole in the trunk.

Back at the house, I try to install the tank onto the stove but can't find the thread.

Calm down, I tell myself.

It's just a flamethrower. Unless I screw up—then it's a bomb.

No pressure.

I try counterclockwise, and the rivets take hold.

Shaolin asks if I have the receipt to give to my school. I tell her I shoved it in a tree outside.

It's worth RMB 100 (about $20), and I decide to lean into hopeful optimism. When I go out later to pick up a food order from the gate, I find it—folded up neatly in the cranny of the tree.

Twenty bucks is twenty bucks.

February 18, 2020 – "An Eerie Journey"

Tuesday – Day 25

Human beings are wet.

We're made of 60% water. The Earth's surface is covered by 72% water. We can live for weeks without food—but only four days without water. Incredibly, we still can't breathe water... and how susceptible we are to pneumonia.

The shocking news of 99 new infections on the *Diamond Princess* is grim. In Cambodia, the Prime Minister shook hands and gave roses as *Westerdam* passengers disembarked.

Let's hope he washed them well after.

Homemade bread with Hainanese honey-kaya and boiled eggs makes a delicious brunch. We get a package at the gate, so I suit up to head out just as my friend Andrea calls. Worried about my cabin fever, he's taking me on a virtual tour of Chongqing's streets.

Andrea leaves his compound, gets his "exit pass" stamped, and walks through the deserted shopping district. He descends into the subway as I arrive at the school gate, where our head of security gives me a wide-eyed stare. Most people wear masks and stop there.

Andrea laughs, but he gets it.

It's a numbers game.

Some people step into traffic without looking up from their phones. Most of the time they're fine—until they're not.

You improve your odds with situational awareness. Wearing a mask in China these days is like looking both ways before crossing the street. Goggles and gloves? That's like listening for invisible cars. But we don't know what we don't know.

The train pulls up. Andrea boards—utterly alone. He stands in the center, careful not to touch anything. From my end, I watch the ghostly passenger in his phone camera, silently drifting through an empty train.

At my gate, I find a small envelope with my name on it—inside are keys from my coworker, Michael. He wants me to loot his flat, grab his laptop, and send him lesson plans. He's offered snacks and water in exchange.

At the transfer station, Andrea stops to take a panoramic shot. Not a single person is visible. It's the equivalent of Grand Central Station for a city of 31 million—and we are completely alone.

On the busier Line 3 train, there are a few scattered passengers, standing far apart or sitting one per bench.

I unlock Michael's apartment. It bears the marks of a hastily packed trip. His laptop is on the couch.

Andrea exits the station and walks past shuttered Prada and Louis Vuitton shops. All foot traffic is now routed through one central entrance. Temperatures are checked, and masks are mandatory. Inside Olé supermarket, they check again.

Back in Michael's kitchen, I grab a big jug of spring water and pack a small box of spices and snacks. He's encouraged me to use them—he won't be back for some time.

Inside Olé, staff are keeping shelves well-stocked. Andrea stops at the beer section and offers to send me a case of Corona. We share a laugh.

A few shoppers wear gloves, filling their carts cautiously. The mood is quiet, orderly—masks on, anxiety down.

I thank Andrea for the tour and return home. I go through my decontamination process, then settle in with my "loot."

Andrea catches a taxi back. At the gate, he gets a temperature check, shows his ID and keys, and is allowed back into his compound.

Today, over 80 clinical trials are ongoing. So far, the front-runner is Chloroquine—**Queen Chlora**.

The annual "two sessions" of China's national legislature and top political advisory body—usually held in March in Beijing—are officially postponed. Chongqing cancels the international marathon scheduled for March 22.

Here in my city, 323 COVID-19 patients are hospitalized—36 severe, 13 critical. We've had 5 deaths, but 225 recovered and discharged.

It's good news, but we're not out of the woods yet.

Shaolin tells me my Chinese has improved. I download books with titles like *Badass Survival Skills*, *Prepper's Survival Medicine Handbook*, and the *SAS Urban Survival Manual*.

It feels good to be prepared.

February 19, 2020 – "Water Under the Fridge"

Wednesday – Day 26

It's water under the fridge.

No, I'm not being irreverent.

My beautiful merino wool slippers are soaked. And without them, I can't keep my toes toasty.

Chongqing shares a similar latitude to Miami (29° to 25°), making it the *furnace of China* for most of the year—but with no central heating, winter is still chilly. Getting these puppies dried up is top priority.

I'm cranky because I tossed and turned all night again—sleepless and anxious—but brunch makes it right. We make pancakes, and I crack open a bottle of *canneberges et pommes coulis*. I've been saving it since we pulled over in Manseau, Quebec, to bargain in a little roadside shop overflowing with all manner of squished, dried, and candied cranberries.

I close my eyes and suddenly I'm sitting on a duffel bag full of cranberries, eating squeaky cheese poutine under a maple tree as the lazy summer wind kisses my face and those plump cranberry bushes sway off into oblivion.

Today, 74,282 people in China have been infected.

14,601 have recovered.

2,009 have died.

In Chongqing, 296 are currently hospitalized. 254 have been discharged. Five are dead.

It's been confirmed that aerosol transmission is possible in closed environments, so we're encouraged to keep our windows open—and the AC off.

I grow these amazing little bean sprouts from seeds. They bring me joy. Next, I'll try green onions, garlic, celery, or lettuce.

A friend calls to tell me our good buddy Simon and his girlfriend are aboard the *Diamond Princess*. I message Simon, but it's old news—he's been back for a while. They're still in touch with friends on board, and it's a total gong show.

Scientists are calling the Japanese quarantine—aka *virus incubator*—an utter failure, but no one can explain how 3,700 people locked in their rooms are spreading infection.

A charming retired couple from the U.K. broadcasts daily updates on YouTube. Today, men in hazmat suits are taping plastic sheets over air vents in the ship's halls. Within a week, the couple is in a Japanese hospital, fighting for their lives.

Watch out for invisible cars.

Meanwhile, the water heater in our bathroom is slowly leaking. I place a bucket underneath. Shaolin is still worried. We're not about to call a repairman over, so we just unplug it.

I love my wife.

She's tough—even by Chongqing standards. Imagine the spiritual wisdom of Yoda, the chemistry of Wayne's-World-era Tia Carrere, and the comedic genius of Mr. Bean.

She's terrific. Even though she pushes me hard.

The Chongqing government releases official back-to-work protocols. They look reasonable—but seem challenging for ordinary people.

Commuting:

Keep a one-meter distance from others. Wear masks and gloves. Avoid rush hour. Drive your own car, if possible.

At the office:

Keep one meter from coworkers. Wear a mask. Disinfect regularly. Work online when you can. Temperature check at the entrance. Use drink containers with lids. Take the stairs. Keep windows open. Wash your hands often.

In the cafeteria:

Wash hands before meals. Dine separately. Bring meals from home. Don't share food.

In the restroom:

Keep a one-meter queue distance. Close the lid before flushing. Wash hands thoroughly.

Since most public toilets here don't have a lid, I'm not sure how realistic that part is—but like good principles, maybe they're meant to be worked toward, even if never fully achieved.

February 20, 2020 – "A Grain of Salt"

Thursday – Day 27

The molecular structure of matter is determined by the chemical bonds between its atoms. Through an electron microscope, cubes appear like ancient monuments, concealing secrets in the recesses of their cubic prisms.

This is a grain of salt.

Instead of listening to the 99th COVID video of the night, I choose a Stoic video on *Memento Mori*, and my anxiety retreats a little. The heaviness eases, just for a while.

Memento Mori is the ancient practice of reflecting on our inevitable death.

Am I really afraid death might rob me of another night sitting on my couch, binging Netflix?

If I'm going to fear death, I should also worry about not living my best life.

Eventually, I sleep for a few hours.

It's harder to be afraid of a virus when you remember that, soon enough, we all will die.

Later, I can't resist the siren call of the news. I listen while I cook and clean, then take notes and write more of my fantasy novel. I keep my nose down and away from the windows.

I know something is out there.

Something terrible.

Invisible.

Deadly.

I can't see it, can't comprehend it—but it's out there.

So we hide inside and hope it doesn't notice us, alone in our building, curtains drawn, drains covered.

I make banana pancakes.

While tidying up, I discover some broken glass on the floor. It slashes open my finger. It bleeds—surprisingly much. A dark red like aged wine, staining the sink. A cloud of my DNA spreads down the drain, tempting the virus to find me.

Finally, the wound clots. No stitches needed.

A super spreader in South Korea infects dozens in Daegu. Now, a city of 2.5 million looks as quiet as any place in China.

Dr. Iwata, a veteran of Ebola and SARS, speaks out about the *Diamond Princess* fiasco:

"It's a COVID-19 factory... I was so scared."

Officials, not doctors, were running the quarantine. Workers ate while wearing contaminated gloves.

He left for self-quarantine.

The passengers waltzed into Japan.

This is going to affect the Olympics, I think. By mid-afternoon, Japan is openly considering postponement or cancellation.

Finger on the pulse.

If we learn anything from this—it's that infectious disease control should be handled by **experts**, not **officials**.

I check my teeth in the mirror and wonder when I'll be able to see a dentist. What an unpleasant, high-risk form of punishment I'm craving.

Speaking of saying *"ah"*, scientists now support the idea that aerosol transmission through ventilation is possible.

After dinner, a boy calls—a friend of my principal—asking for help with a speech to commemorate the 70th anniversary of the People's Republic of China. He dreams of winning a city-wide contest, remixing the national anthem, and leading China to power and glory. I correct his grammar and pronunciation, and he's a happy little guy.

According to the Chongqing Economy and Informatization Commission, most of Chongqing's key companies have resumed production.

I've also heard that hand sanitizer stations are springing up all over town.

I've applied to test the new *Cyberpunk 2077* game and might take a job writing for another one. I volunteer some time for my favorite progressive candidate's campaign.

These feel like good ways to spend time.

We need a Green New Deal more than ever.

I have a song stuck in my head.

I don't know why.

Every day is exactly the same.

February 21, 2020 – "What a Time to Be Alive"

Friday – Day 28

I'm in a basement apartment, trying to dispose of a pile of dead bodies. We find a wood chipper under a tarp and fire it up. I choke back a scream as it roars to life. We're about halfway through when the neighbors start banging on the door, so I yell to Shaolin over the deafening engine:

"Put the kettle on!"

I wake up early, in a cold sweat.

What was that?

Horrible nightmares.

A live COVID-19 Q&A begins with Dr. John Campbell. More than 4,000 of us are tuned in. Despite his cough and cold, he patiently answers questions for hours. I make some coffee and check in with my daily dose of viral pathologist Dr. Chris Martenson.

I help Shaolin apply medicine to her sore shoulder.

The new numbers inside China are low, and many companies in Chongqing are back to work.

Fantastic.

Two of the most severe cases from the *Diamond Princess* in Yokohama turn out to be Japanese government officials. The ship ultimately reports 634 confirmed infections—starting from a single passenger who boarded in Yokohama and disembarked in Hong Kong.

In Canada, British Columbia reports a new case linked to Iran's emerging cluster—shifting the spotlight from China to a more global spread.

My school asks my colleagues to return for a 14-day quarantine, while Canada, the U.S., and the U.K. embassies are still advising foreign nationals to leave China. I'd prefer to continue teaching online until restrictions on travel and public gatherings are lifted.

If a beer at the pub and a movie at the mall aren't safe, how can a class-room be?

The fridge seal is loose. I have a strange cramp in my left leg.

Entropy is a force without mercy.

I rub on some Arnica and stretch. I used to leg press 300 kg. Now my calves look small and soft. Still, I'm another kilo lighter today—just a few more to go by summer.

I suit up and find our number printed on a big white cooler wrapped in plastic at the gate. On the way back, I tear into the packaging and re-open the cut on my finger.

It's fish.

Later, I try navigating the shower with my eyes closed. The practice has three goals:

1. Ease the fear of losing my sight.
2. Make me more grateful that I *can* still see.
3. Avoid thinking about virus particles splashing into my eyes while I wash my hair.

Shaolin spends two hours cleaning and preparing a beautiful dinner.

By yesterday, the super spreader in Daegu, South Korea had infected close to 100 people, all linked to a religious retreat. Millions in Daegu are now hiding from the virus too.

The coronavirus can travel across the world in a day, as fast as a plane can carry it—

but fear still moves faster.

Later, I get a code for a free PS4 game. The devs want to hire a writer and ask me to play it and write them some stories. I'm exploring the game when Shaolin comes in, asking me to prep for class.

I hold up my controller and say:

"I'm busy working."

One of those moments where we don't quite understand each other.

I apply for a job on a remote island community in Ireland—they're looking for a cheerful couple to run a coffee shop in exchange for free housing. It's fun to dream. I'm halfway between craving a wild night out in the city... and wanting to buy a little cottage in Quebec and live off the land.

Close to midnight, I get a package notification. I suit up again. Shaolin hasn't left the flat in two weeks. I'm more comfortable with our protocols, while she'd rather not take chances.

Through the foggy darkness beyond my goggles, I see flashes of light.

And I hear the screech of violin strings.

It sounds like the score of a horror movie.

No one is around.

It's terrifying.

What a time to be alive.

February 22, 2020 – "We Rise Again"

Saturday – Day 29

I sleep well and feel good today.

We teach a morning class while laughter and the grinding of coffee beans hum in the air. For lunch, we make our famous spicy Chongqing noodles. One day, maybe we'll move back to Canada and open a restaurant—or a B&B by the water somewhere scenic.

In Ukraine, frightened protesters clash with police, trying to stop passengers returning from Wuhan. The panic fueled by social media can be as dangerous as the virus itself.

In Italy and Iran, people wear masks in public as the infection spreads. Many Korean neighborhoods now look as empty as Chinese streets.

A Canadian woman who contracted COVID-19 in Iran triggers a "sentinel event," broadening Canada's and the CDC's criteria from "Have you been to China or know someone who has?" to reflect a new global climate. Now, with nearly 30 countries seeing domestic transmission, potential cases could come from anywhere.

Increased precautions are essential, and frontline workers must prepare. After lunch, I organize my digital life and read a little. We teach another class in the afternoon, and make an early dinner: fried potatoes, fish, and green veggies.

One last lesson in the evening, then it's time to relax.

Keeping busy makes the day fly by in a comforting way.

I've decided to limit my COVID-19 news consumption to daylight hours—and wind down with some self-care in the evening.

In China, President Xi Jinping says the turning point still hasn't come and that the situation remains severe.

This comforts me.

We all want it to be over—but this thing has to burn itself out. If we celebrate too early, the quarantine will have been for nothing. As long as I can see my grandma in Canada this summer, I can handle anything else.

Packages arrive—dog food, coffee, and avocados.

I talk to my good friend Stu. He gives me a real boost. He's positive, cool, and insightful.

The things we've seen since we were kids still astound us.

What a time to be alive.

I grew up when kids played outside. I witnessed the birth of the cellphone—from bricks to flips to sleek smartphones.

I watched video games transform from blinking pixels to immersive virtual reality, indistinguishable from life.

For the next thousand years, the digital age will define humanity—but I saw the beginning.

I remember dialing up directly into bulletin boards before the internet was *the internet.*

I watched globalization go from abstract concept to seamless, AI-managed reality.

Global transportation became instant, affordable, and reliable for billions.

Dance music went from underground warehouse parties to stadium shows and TV ads.

I lived through the height of Western cultural influence—and then moved to China as it rose to stand beside America in terms of power and global significance.

Automation and artificial intelligence now reshape our world daily.

This virus may disrupt the chain of global movement.

But it will also force us to become more resilient.

We've gone through worse.

And we'll rise again.

February 23, 2020 – "Patience Prevails"

Sunday – Day 30

The word *quarantine* comes from the Venetian form of the Italian *quaranta giorni*, meaning "40 days." In the old days, sailors had to dock and wait forty days to prove they weren't carrying disease—often while rats onboard snuck ashore and spread the plague anyway.

I've got ten more days to go before a real quarantine is technically complete.

I close my eyes and breathe in the memory of endless lavender fields in Provence, brushing bees away from my feet. We spent Shaolin's birthday sailing around a volcano off the coast of Santorini, then marched those merciless ancient Roman cobblestones that broke her suitcase and tore her shoulder.

I remember seeing an old woman face-plant in the middle of a crowded piazza that day. She dropped face-first into a busy crosswalk as her sweaty, bewildered family struggled to lift her up. Her nose poured a fountain of blood onto the old stones—like something out of Fellini. A proper Italian *"Ma che vuò?"* of nosebleeds.

Europe is still in our rearview mirror, but it also feels like a lifetime ago. We relax for a while. I help Shaolin with her shoulder treatment and make a breakfast of oatmeal, boiled eggs, hummus, and toast.

We teach a good class—and then a hard one. The audio lag is massive. Usually, I'm asking students to *take* their earbuds out; now I'm begging them to *put* them in. I mentally check out and let Shaolin take the lead. She's convinced I've been cursing under my breath.

Unless she can read minds, she's mistaken.

Later, she stomps around the kitchen, banging pans.

Arguing during a global pandemic is high-stakes—you can't just take a walk to cool off.

So... patience prevails.

In a wild fantasy, I Google flights. Thailand to Canada, one of the only direct routes still running from Chongqing. South Korea has already issued a travel warning for Bangkok. Thailand might issue one for South Korea tomorrow.

It's expensive anyway.

Jeffy Spaghetti.

I revisit my writing community on Scribophile and start a new group for locals:

Cyber Chongqing & the Hot Po(e)t Society.

My buddy falls over and messes up his knee. He wants to go to the hospital but he's scared.

He asks me what to do.

I recommend ice, heat... and prayer?

I re-refry refried beans with onions and garlic, then make a glorious toasty burrito stuffed with a fried egg, hummus, salsa, cheese, and hot sauce.

I'll remember this burrito forever.

I wish I had some beer.

Today is cold.

I'm excited for warmer weather on the horizon.

I have a long call with Andrea, and we laugh until my gut hurts—at the absurdity of it all.

And yeah, I've got a tough wife.

You know who else did?

Shakespeare.

Lovecraft.

All the world's a stage—

Full of cosmic horror.

February 24, 2020 – "Good Housekeeping"

Monday – Day 31

Normalcy bias is the human tendency to believe things will always function as they always have. It causes us to underestimate the probability—and consequences—of disaster.

We see it in discussions about the environment, sustainability, animal rights, and of course, infectious disease.

I prepare for online teaching with a coffee that reflects my heart:

Strong and black.

I send a few friends photos of gutted Italian supermarkets.

Now might be a good time to stock up on a few months' worth of rice and medications.

Being prepared is one of the few things within our control—if we're willing to break through the fog of normalcy bias.

The first step?

Situational awareness. Trust your instincts.

Today, the first chartered train for 500 migrant workers from 30 districts in Chongqing departs for Zhejiang Province.

Chongqing reports two new infections today.

There are currently 234 hospitalized—21 severe, 10 critical.

We've had six deaths.

335 have recovered.

I spend the morning trying to convince Shaolin to order more rice.

Three packages arrive, but no rice.

I gear up and crank my rock cover of "My Corona" in protest.

We've got weeks' worth of greens, milk, and yogurt...

But no rice.

Shaolin shrugs. Her favorite Thai rice was out of stock.

I'm not impressed.

Even if I think this will all be over soon, I start boiling water to fill our empty spring water bottles. Two liters a day per person means I can store a month's worth, just in case.

I start a new diet plan: **one meal a day.**

Fasting for 23 hours helps your body clean up dead cells and strengthens the immune system.

My massive lunch includes:

- Three boiled eggs
- Refried beans
- Quinoa
- Toast
- Hummus, salsa, and hot sauce

It's hard to finish, but by sundown my belly is rumbling again.

My slippers are sticking to the floor.

I miss our friendly, affordable, and familial *ayi*—our housekeeper.

For six years she kept this place spotless.

Now she's in her village, and I spend two hours mopping, almost slipping half a dozen times.

I repurpose a magnetic strip to fix the fridge door.

Then Ben Ben poops all over the living room floor.

As I banish him to the balcony, Hachoo locks eyes with me in full solidarity... and pees a massive puddle.

I scream until my throat burns.

Then I mop the floors all over again.

That's just good housekeeping.

February 25, 2020 – "Passion, Purpose, Progress"

Tuesday – Day 32

I stay up half the night, sending out hopeful messages and lists of emergency supplies to family and friends. COVID-19 is coming to their communities, and I hope they'll take precautions.

For many, it's still not even on their radar—even with the shift in media tone and growing talk of strict quarantine measures, like those now implemented in Italy's posh northern Lombardy region.

Shaolin wakes up coughing, and I'm quietly concerned.

She's been a little withdrawn and grumpy since Sunday, but she gets up and makes pancakes.

I enjoy them with yogurt and coffee.

We binge the entire season of *Locke & Key*—Joe Hill's show. He's the spitting image of his dad, and the series is excellent. I alternate between work and exercise while watching.

In absolutely fantastic news:

China enacts a ban on the wild animal trade and consumption, long suspected to be behind both the SARS outbreak and now COVID-19. It's about time.

I just hope it lasts.

Without a market to sell to, maybe the world will be a little safer—for our tigers, sharks, and bears (oh my!)—today and tomorrow.

Meanwhile, my dogs are protesting again.

Peeing and pooping on the floor.

I wish I could explain that it's not safe for them outside—for a whole host of reasons. Not just infection risk, but because frightened locals might see them as health hazards.

My colleagues in the U.K. and Germany reach out, asking about our health insurance and pay situation. Our countries are still warning us to return home. There have only been three new cases here in the past two days—but one is close to home: a man who works in a butcher shop five minutes from my school is now infected and hospitalized.

Today I "blow" my one-meal-a-day diet while helping Shaolin cook dinner.

We end up eating together—a simple meal of greens and rice.

I can't justify skipping fresh vegetables, especially when they're this good.

Due to local hotels being full, China is now sending **seven cruise ships** to Wuhan to house the wave of healthcare workers arriving from all over the country to help treat the infected at the epicenter.

Around 9 p.m., I steal an Oreo.

Because dieting is hard.

I watch some of Apple's *Mythic Quest*, a comedy about game developers, and find myself wondering what it would be like to work for a Chongqing-based video game company. I think I'd enjoy it. I plan to write them some great stories and see where it goes.

Shaolin naps after dinner, and I settle into a book, knowing I'll go to bed early too.

This has been going on so long that I'm no longer glued to every breaking headline.

For the first time in a while, I feel like I might actually be able to relax.

February 26, 2020 – "A Piece of Cake"

Wednesday – Day 33

I wake up at 11:11.

Make a wish.

Old friends I haven't spoken to in years are reaching out. I've got lots of fun things I want to do today, but I start with the news and a strong cup of coffee.

I'm told I need to collect more invoices from a tax office if I want to keep getting paid. I ask them to just hold onto my money.

I'm surprised when a delivery guy is allowed to come all the way to our building. I pick up the package and Shaolin tells me things are almost back to normal.

I'm not so sure.

We practice being independent of society's daily interactions, living in isolation. But that insulation doesn't come free from risk.

A woman in Wuhan demonstrated how deadly chloroquine can be when taken improperly. She ordered the drug online after suspecting she'd contracted the coronavirus. She hadn't. But the 1.8-gram dose she took caused a malignant cardiac arrhythmia that landed her in intensive care.

Then Donald Trump tweets about chloroquine—*Queen Chlora*—and some Nigerians overdose on it, too.

They almost die.

I work on a new song on my ukulele and sign up for an MIT course called *Principles and Practice of Human Pathology*. To boost my productivity, I subscribe to a lecture series called *Pathoma*.

Later, I take a break and get lost in the pages of a juicy sci-fi novel.

Still, I worry.

People are starting to come back.

What happens if they bring it with them?

What will China do—restrict reentry?

To relax, I paint Dungeons & Dragons miniatures under the warm light streaming through my window. My shoulders relax, and for a moment, I'm 12 again—shopping with my dad for miniature heroes and monsters.

I used to be good at painting.

When I picked it back up, I had to resist the urge to glob paint everywhere.

Instead, I relearned patience—how to dry brush, how to layer.

Discipline and restraint.

Being okay with making mistakes.

These are grounding techniques. They make room for creativity and joy.

Shaolin tries to buy some vitamins, but they're stuck—parked somewhere on the Yangtze River at Wuhan Port.

We agree: this summer we'll go to Canada and stock up.

Part of me dreams of buying a little cottage and becoming an off-the-grid hermit.

But once this passes... living in a high-tech supercity is still pretty cool.

I make pizza sticks with a wasabi hummus dip—

A fun little invention so refreshing it brings tears to my eyes.

We have noodles for dinner, and Shaolin wants to try making an egg cake.

I'm enthusiastic in my support and help her in silence,

because she's still really pissed at me.

It works out beautifully.

She's delighted with it.

Sometimes, when life cracks your eggs...

You just have to make a cake.

February 27, 2020 – "Shopping with Braveheart"

Thursday – Day 34

Bravery is not the absence of fear—it's managing anxiety and holding steady when your body tells you to shut down or run.

I've always considered myself brave, but I'm rattled.

I've never endured this much constant, low-grade existential stress for more than a month—without finding some kind of "out." I need a new approach.

A few hundred years ago, in the early 18th century, one of my ancestors—Rob Roy MacGregor—made a name for himself defending his Scottish land from invaders. He became a folk hero. A real-life "Scottish Robin Hood."

I'll channel that bravery today.

Get off my land, you invisible parasite.

Today, we lose another hero—Dr. Li Wenliang.

He warned the world. He said from his hospital bed that his "mistake" was examining an elderly asymptomatic patient with his mask off.

It was only for a moment.

Two weeks later, from ICU, he tweeted that he hoped to recover.

But he didn't.

Don't believe the lie that this virus only targets old people.

I've been doing a lot of outreach on social media—trying to help friends and family prepare. It's exhausting, but not doing it would be worse if something happened to them.

When people tell me I'm brave, I laugh.

Putting one foot in front of the other doesn't feel heroic.

We're all just trying to survive.

I'm scared for my family back in Canada—still out and about, exposed.

Last night I stayed up until dawn reading medical studies.

Even if China is stabilizing, paranoia clings to me. I need to be prepared. I found an article in *Maclean's* about Dr. Michel Chrétien, a respected Canadian scientist (and brother of former Prime Minister Jean Chrétien), who's testing high doses of Quercetin—a plant-based supplement previously studied for SARS and Ebola. He's heading to China for human trials.

At 5 a.m., I'm under the covers, driving Shaolin crazy while searching for vitamin shops. I find one. They've got four bottles of Quercetin left. I buy them all.

I wake up exhausted and depleted.

We eat egg cake and drink coffee.

I have a meeting with Liz in Toronto and post some things for Democrats Abroad.

I believe in politicians who fight for people.

When I hear young people respond to a pandemic with, *"Good on you, Mother Nature,"* I know in my bones—it's time for a **Green New Deal**.

Today marks the first day there are more new cases **outside** China than inside. That feels significant.

Online, most people are still saying,

"Oh yeah, it's just the flu,"

or

"I'm not old, I'll be fine."

But this doesn't feel like the flu.

I've never holed up in my house and seen schools close for a month because of the flu.

I weigh myself—down another two kilograms.

Shaolin's salsa teacher Luis from Havana messages me on WeChat: "Are you losing weight?"

Maybe I'm born with it. Maybe it's COVID-19.

I'm watching Iran closely. Their unusually high death rate for confirmed cases suggests either a particularly nasty mutation—or thousands of undetected cases. A Canadian statistical model estimates there are actually **18,000 cases** in Iran right now. They're not canceling public events, but neighboring countries are closing borders and canceling flights.

We can't order more rice, so I make the call:

It's time to **suit up** and go shopping.

The streets are mostly empty.

I pass a public toilet and hold my breath to avoid aerosols.

For once, I manage to adjust my goggles so they don't fog in the first five minutes. The supermarket is quiet.

But as shopping drags on, I start to feel *exposed*.

There's this panic—this *feeling*—like I should keep moving, like I'm breathing too much.

I don't like it.

Shaolin's coaching me via video chat.

Then an older man behind me starts coughing—violently.

I bolt down the biscuit aisle with my cart.

The vegetables are well stocked, but mushrooms are gone. I find no sugar in the usual section and scrape the bottom of the barrel in bulk.

I get a big bag of rice, four bags of Doritos, lots of veggies, pasta, and some meat for Shaolin.

It's all expensive. But worth it.

On the way back, I struggle with four heavy bags. They've cordoned off part of the sidewalk and make me cross the street. I stop, strip a few layers, and slow my breathing. I **must not** fail the temperature check and get dragged off to the fever house.

At home, I'm sweaty and out of breath.

I toss my clothes in the washer and hop in the shower.

I forgot to plug in the water heater.

Iceman style.

I do some laundry and make a tomato and cheese sandwich with wasabi mayo. Later, I have shrimp wonton soup but my throat is a bit icky.

I suck on a few *Fisherman's Friends* from the care package my mom sent and take a nap.

Later, I drink honey-echinacea tea and we watch some movies.

Tomorrow, I'm going to make French toast.

And ignore the world.

February 28, 2020 – "The Winds of Change"

Friday – Day 35

I wake up and kiss my wife.

She kisses me back.

All traces of anger, gone.

Every fire runs out of fuel.

"Don't panic!" cries the internet.

Can anyone check if we still have dolphins?

My electric toothbrush finally died.

It was a gift from my stepmom, Ming, part of a back-to-China care package after a long, wonderful summer vacation in North America.

Now I've got a new toothbrush, plucked from our stockpile of provisions.

One day you have one thing, and the next—another.

Don't feel sorry for the toothbrush.

It's just a toothbrush.

I loathe to mention it, but Hong Kong just reported an infected dog.

There's no evidence it'll get sick or pass it back to humans—but fear spreads faster than COVID-19.

Keep your dogs inside.

My boss has promised me a glowing reference letter for six years of great work. Today, I write one for the most promising student I've had in all those years. I hope she lives an amazing life in college and beyond.

I go to pick up our packages.

My winter coat isn't going to cut it anymore.

In minutes, I'm drenched in sweat, breathless, and panicking.

A guard at the gate passes his electronic thermometer to a passenger in a parked car. She's rubbing her unmasked nose.

She hands it to the driver, who shoots himself in the head—figuratively—and then passes it back to the guard.

That's how easily you can get COVID-19.

I'm frantic, scavenging for our packages in a sea of cardboard.

Shaolin tells me to calm down.

But I can't get a grip.

Eventually, I find them.

Now we have apples.

Brunch is avocado toast again. I don't stop to take a photo.

The sunbeams tease us through the windows, so we suit up and head out for some air.

We sit out on the garage rooftop in lawn chairs and just breathe.

It's Shaolin's first time outside the apartment in weeks.

I take off my jacket, gloves, and mask, and slip on my handcrafted Italian Ray-Bans.

In a T-shirt, I soak up the vitamin D.

My skin tingles.

Today, Chongqing is 18°C and sunny.

The sky is baby blue, the clouds fluffy and slow.

It feels like spring.

Over the past month, I've researched and learned, fanned the signal fires and prepared. I've cycled through at least five stages of grief and adaptation.

Now I'm ready to get on with it.

We will move forward, one foot in front of the other.

It will be hard.

We won't all make it.

But most of us will.

I really believe that.

I used to be the kind of guy who'd crack a can of soda without washing the top.

Now I know things I can't un-know.

All the world's a stage—
Full of **cosmic horror.**
And yet—
I feel good.
I've had an incredible life so far.
And I, for one, didn't survive the '90s rave scene and two decades as a touring performer
just to be taken out by a virus named after a light beer.
I've got a fire in my belly.

PART III MEANINGFUL ENGAGEMENT

February 29, 2020 – "Leap Year"

Saturday – Day 36

It's a Leap Year.

And what a lucky thing it is to have an extra day.

Even today.

I wake up at 9 a.m. and brew a strong, robust Italian coffee.

I teach a fun class at 10 a.m., and while I'm making brunch, Shaolin suits up and heads out alone to retrieve a package.

I'm shocked.

I feel like a worried dad waiting for his teenager to come home from their first date.

She returns with our stuff, calm as anything.

Weeks of not wanting to leave the apartment, and now—she's found her legs.

As the wheels of life begin turning again, a quiet relief washes over me.

I'm proud we rationed carefully and stocked up on coffee, water, eggs, rice, and other essentials.

The supply line never broke.

Chongqing took extraordinary measures to keep supermarkets well-stocked and stable.

My precautions weren't necessary—but I don't regret them.

As sunlight spills through our windows, we let the dogs sunbathe on their cushions. We change their heavy pajamas for lighter ones.

Technically spring begins March 20, but today feels like a cool spring day already.

They say Chongqing has only two seasons: summer and winter.

For most of the year, this place burns at 40°C.

Today it's 18°C and sunny, and the sky is baby blue.

Shaolin and I are working well together.

We teach another class from 2–4 p.m., and I make a playlist called *Covid Choons*.

Track one, with a bullet, is "Guided Explorations," a meditative release
by RZA & Tazo.

I listen to it on repeat until I have it memorized.

It calms my nerves. I feel the weight dissolve from my shoulders.

I'm not so snappy anymore.

RZA teaches me patience.

Shaolin teaches me love.

And the kids we teach are happy.

I talk to many friends. Some are receptive, ready.

Others are confused, dismissive.

They ask if I'm drunk or going through something dark.

They just don't see what's coming.

But a few get the message.

And that's all I can do.

Today, again, the number of new COVID-19 cases **outside** China ex-
ceeds those **inside.**

SARS-CoV-2 is on every continent except Antarctica.

Sixty countries today.

We'll hit millions of cases by April if we don't stay home, distance ourselves, and start connecting inward instead of outward.

New cases are reported in Denmark, UAE, Azerbaijan, Iceland, Lithua-
nia, Mexico, Nigeria, and Britain.

France, Italy, Iran, South Korea, and Japan are now *popping* cases like
Jiffy Pop on a stovetop.

In the U.S., the case counts don't add up.

Some call it *"Don't test, don't tell."*

A whistleblower reveals that quarantine workers weren't even tested before boarding commercial flights home.

We cannot afford amateur hour.

Economists across the spectrum warn that a crashing market could cost
Trump his re-election.

Some say it already has.

Shaolin and I have a nice Chinese-style dinner—plenty of greens and mountains of rice.

But our final class of the day wears me down. The buffering, the fidgeting, the blank stares.

I'm trying to teach advanced concepts to kids who'd rather be playing video games.

But I get through it.

A sobering call with my dad paints a grim picture back home.

Ottawa's Costco is handing out Lysol wipes at the door.

Some people argue rather than clean their hands.

Shelves are stripped bare—water, food, TP.

Shoppers wearing masks stop to sample tiny pizza bagels.

Bless their hearts.

Italy, already Europe's slowest-growing economy, is now excluding asymptomatic positive test results from its totals.

Europeans are price-gouging each other for masks.

Everywhere has shortages—except the Asian countries that actually produce them.

In the U.S., the head of the CDC testifies in Congress: *"We don't recommend prepping."*

Their website tells a different story—warning people to mentally prepare for closed schools, canceled events, self-isolation, furloughed jobs.

We snack on spicy local Chongqing snacks and watch some TV.

It's a lazy afternoon.

When we tune out the chaos, we can still feel happy.

The stock market continues to slide.

Analysts warn of a free-fall.

Supply chains are breaking.

Buy what you need—especially for colds, fevers, and headaches.

Unless you're critically ill, you will want to stay home.

And then—this headline leaves me gasping for air:

"Coronavirus outbreak at cyber goth rave kills zero."

At least the burners, cyber goths, and industrial crowd have the gear.

I always knew Burning Man was training camp for societal collapse.

Midday, I switch gears from doomscrolling to *Critical Role*, a four-hour D&D podcast.

I'm going to start an online game with friends soon—something creative to tether me to joy.

Painting. Reading. Music. Play.

At night, we watch *Carriers*.

It hits different.

Feels like 5D in this climate.

Before bed, we switch to *Blades of Glory* and laugh our guts out.

Laughter is good medicine.

March 1, 2020 – "Beans, Bullets and Bandaids"

Sunday – Day 37

Three days until my quarantine party.

It started as a joke this morning but, by evening, had refined itself into a focus-tested plan.

Today, it's 21°C and sunny. A true *T-shirt kind of day.*

The air is clean. The streets are quiet.

A silver lining: fewer cars on the road, a rare break for the environment.

"A Coronavirus Movie," the popular Twitter account narrating the pandemic like a Hollywood thriller, retweets my blog—cementing my role as the eccentric journalist yelling from the rooftops to stock up on beans, bullets, and Band-Aids.

Ben Ben looks at me, confused and a little hurt.

Why don't we go outside anymore?

Why don't we play like we used to?

He's happy we're spending so much time together in his old age.

He just doesn't understand why I won't open the door.

Shaolin's flowers are stunning today—deep lavender, royal blue.

They bloom slowly but persistently on the balcony and windowsill, rising against gravity, a small rebellion.

A *Will to Be.*

Fragrant. Bold.

Alive.

I make a tuna sandwich with my custom handmade Damascus steel blade.

I'm a *stuff* guy.

Shaolin and some of my friends call me a hoarder.

Collector. That's the word.

The cucumber is crisp, the toast golden, the tuna reminds me of Gatineau Park picnics when I was young.

Four hours of teaching today.

I do my best to ignore the distractions my own mind invents.

It goes better than last week, mostly thanks to my hip-hop meditation routine.

Sometimes I slow down so much I feel like I might vibrate into another dimension.

But I hang on.

It cements a truth:

I'm as flawed as a sidewalk—cracked, uneven, yet capable of blooming the occasional daisy.

Like an old willow tree, stubbornly reaching toward the light.

Many of my writer friends are producing new work.

Quarantine has been good for creativity.

I take a moment to #MeToo myself—it's 2020.

You don't grow up in the '90s and land in 2020 without realizing that "boys being boys" was actually pretty uncool sometimes.

I try to be a good husband.

I try to be a good teacher.

I stumble across an insane list of survival gear online.

If Shaolin wouldn't leave me, I'd buy half the site.

(Just kidding, Shaolin... mostly.)

I'm eyeing a solar generator—because if the grid goes down, I'm not losing my music.

A UV lightbox to sanitize masks.

I'm grateful I'm not using cash. Paper money in a pandemic? No thanks.

We start a video class.

Shaolin calls, "Lucy... Lucy...!"

I close my eyes and time-travel back two years:

"Lucy... Luuucy, you got some 'splainin' to dooo...!"

Everyone laughed—even Lucy.

They didn't know who Ricky Ricardo was. But it was funny anyway.

Lately, I've been time-traveling more.

Dreams of Canada feel real, like parallel lives still unfolding.

Maybe time isn't linear.

Maybe it's a Möbius strip.

If I squint just so, I'm 18 again.

It's 1999.

And I'm prepping for my first international gig in Brooklyn, New York.

This year I finally replaced my old Road-Warrior MacBook power adapter with a tiny, dainty cord.

Then this happened.

The world isn't ready for dainty right now.

Back to basics.

"If the coronavirus doesn't take you out, can I?" texts Dave Mile, eyebrow emoji implied.

China represents 30% of global manufacturing in 2020—more than double what it was during SARS.

When we shut down for even a month or two, global supply chains suffer: medicine, car parts, electronics.

We don't save. We don't store. We optimize.

Global capitalism is lean—but fragile.

Now economists are melting like gingerbread men in the rain.

They used to call a mild economic dip a *"haircut."*

A China-first-quarter haircut. A U.S. one soon to follow.

Me? It's been two months since my last haircut.

Misinformation is costing lives.

In California, 100 frontline workers are self-isolating because the CDC misclassified a patient as droplet instead of aerosol.

The CDC. *Bird's eye view?* We're not ready.

Iran's former ambassador to the Vatican has died.

Pope Francis cancels three days of events with a "cold."

His staff denies the rumors. But we all wonder.

During our classes, I move, stretch, add plates to the weight bar, sneak off-camera reps.

It feels good to move.

I'm a sitting duck, but I refuse to stagnate.

Some professionals are still telling the public not to wear masks.

They say they're "ineffective," yet in the same breath ask we leave them for healthcare workers.

The logic collapses under the weight of contradiction.

In the I.T. world, we call it PEBKAC:

Problem Exists Between Keyboard and Chair.

It's different in Asia.

Masks are cultural. Normal.

Not political.

They worked in 1918. They work now.

I listen to preparedness expert James Wesley Rawles.

He maps out the domino effect:

First we lose masks, gloves, sanitizer.

Then meds.

Then non-perishables.

Then cleaning products.

Then perishables.

He's not wrong. I've seen the photos.

In China, we've been lucky.

A few days without mushrooms. A sugar shortage.

But always greens.

Always TP.

People ask, "Do you still have power? Food?"

They're surprised when I say,

It's just like normal life.

We stay home.

We wear masks.

We work online.

And yes, we still have power.

Some of my friends in Toronto report gunshots in the toilet paper aisle.

You can't make this shit up.

Season two of *Altered Carbon* dropped.

Looks great—but Shaolin's not feeling it.

She watches a touching documentary about Chongqing's *bang bang men*—the last porters in China, hauling groceries and televisions on bamboo poles for a few bucks.

She looks up from her phone and says:

"The balcony smells like something died out there. Can you clean it?"

I nod.

Keep typing.

Then she asks me to pick up my clothes.

Rewash the dishes.

Tidy my office.

"I've gotta write," I say.

"Oh, write?" she retorts. "You're a writer now? Where's all your books, *Mr. Writer Man*?"

"Hey, I'm here every day. Is it my fault if I've got a bum muse?"

She shakes her head.

"You're lazy," she says.

Maybe I am.

Or maybe I'm resting.

I wash the dog pee off the balcony.

Before bed, I drink my golden milk.

Haldi Doodh.

Turmeric, black pepper, honey, warm milk.

An Ayurvedic remedy—balancing for all doshas.

Anti-inflammatory. Antiseptic. Antimicrobial.

A comfort. A ritual.

A slow sip of ancient wisdom.

I sleep well.
Even with a yellow tongue.

March 2, 2020 – "Fender Bender"

Monday – Day 38

After a strong pot of coffee, I feel like a jacked-up workhorse.

Memento Mori. Carpe Diem.

I write a glowing letter of recommendation for the most talented student I've had in six years, helping her apply to Stanford.

It feels good to think forward—to a time *after* COVID-19.

To goals. Dreams. Possibilities.

I read the reference letter my old boss Judy wrote for me.

It's beautiful.

Our six years working together were some of the best.

There was a time—pre-COVID—when a new boss felt like a big change.

Now? *Perspective is everything.*

Tom Cruise cancels the new *Mission: Impossible* movie.

He's fine with skydiving, but he's not crazy enough to breathe Italian air.

That says something.

France closes the Louvre to protect the Mona Lisa.

From *us.*

I email Aunt Elaine and Uncle Larry in Chicago, checking in, sending them my blog, research, and the latest draft of my book.

Elaine says they're doing great.

Uncle Howie's in Arizona, visiting Uncle Vic.

My dad thinks he should head home. I disagree.

If you're going to hunker down, Arizona's as good a place as any.

I download a class video about *risk-taking* and send out more teaching materials.

It's a great session—*crisp, clear, perfect.*

And yes, I'm now the second person in history to say that awkward sentence.

Later, Andrea and I play *Shadowrun* online.

I'm a Navajo Ork sniper, a badass gunslinger in 2058 Seattle.

We're prepping for a run.

Of course, everything's about to go sideways.

Preparation is everything these days.

I send Andrea some IELTS prep books—his wife is studying.

They send me Vitamin D3 in return.

Science says D3 boosts the immune system and helps the body resist infection.

I'll take it.

After class, Shaolin and I do a whirlwind clean-up.

Living and working from home now just feels like... life after forty days.

Shaolin tells me to rush to Building Eight to pick up a package before they start charging storage fees.

It's in a signal dead zone, so paying through the app is tough.

Add foggy goggles, sweaty gloves, and a stifling mask—*adventure mode activated.*

On the way up, I pass a fender bender.

A car's smashed into a minivan.

Two families are huddled beside the wreck, masks off, calling for help.

I see oil dripping from the twisted metal—and suddenly, the scene warps.

The lazy heat shimmer becomes a mirage.

I'm no longer in Chongqing.

The school vanishes.

I'm in the California desert, somewhere past Klamath.

A busted radiator hisses in the midday sun.

I'm traveling with Lumo and Chelly in a road-worn van, en route to Black Rock City—a mythical, temporary bazaar of creatives and cosmic weirdos beneath a blood moon.

We've broken down.

The vehicle is steaming.

The desert is endless.

I look down and see I'm standing on the sun-bleached corpse of an earth goat—my spirit animal.

A warning.

I remember this.

The year I became a detective.

More than anything...

The year I found myself.

Then I blink.

Back in Chongqing.

I snap a photo, grab my package, and head home.

But there are *more* packages waiting at the gate.

I suit up again.

We now have tofu jerky, nuts, and a big box of plump oranges.

Later, after hours of procrastination, I steam chickpeas and make **exceptional hummus.**

A proper hummusmancer is born.

Do I still need all these precautions?

Maybe not.

But good habits matter.

If there's a second wave, we'll be ready.

Tomorrow night: D&D.

Wednesday morning: my **40-day livestream quarantine party** on Facebook.

My week is looking good.

I make pasta.

It's delicious.

The local COVID-19 numbers bring good news.

Only 120 confirmed cases still hospitalized in Chongqing.

Nine are severe, two critical.

Six deaths total.

But 450 people have recovered and been discharged.

There have been *no new confirmed cases for six days straight.*

The Municipal Health Commission has tracked 23,563 close contacts.
Almost all have been released from observation.
Only 371 remain.
This is how you do it.
Factories and offices are resuming work.
But schools are still holding back.
Shaolin thinks our fridge door doesn't close properly because of ice.
I think the ice builds up *because* it doesn't close properly.
Solution: I sneak up on it with a hammer, pull out the shelves, and smash the ice to bits.
Result:
Door closes like new.
Sometimes the best answer is low-tech.
Even NASA has the Mars rover hit itself with a shovel now and then.
Later, we relax and watch TV.
Tomorrow is **Dungeons & Dragons.**
And as Matt Mercer says:
"How do you want to do this?"

Fail to Plan, Plan to Fail

March 3 - Tuesday

COVID-19 is now in **70 countries**, with more than **92,000 people infected.**

It's been raining for days. Heavy, cathartic rain. The air smells scrubbed and new.

I pick up two packages: **five pounds of bean sprout seeds** and **organic apple cider vinegar.**

I had asked Shaolin to order the vinegar. She rolled her eyes and gave me a five-minute monologue about how my "weird food is weird"... and then she ordered it anyway.

From my quiet corner of the world, it feels like China is getting things under control.

Even **WHO Director-General Tedros Adhanom Ghebreyesus** said Monday:

"In the last 24 hours, there were almost nine times more cases reported outside China than inside China."

Quarantine works. This shift in the tide proves it.

I fantasize about walking through a mall, watching a movie, sipping a coffee on the cliffs of **Hongyadong**, watching the sun sink into the river.

Soon. Chongqing is waiting to be rediscovered.

We make delicious pancakes.

Later, I brew a strong pot of coffee and do a tutoring class.

Eight-year-old Kim shouts, "Hi fishy, nice to eat you!" and I laugh hard. Laughter is medicine.

After class, I scroll through shopping apps.

Should I order more **immunity boosters?**

Maybe a **P100 respirator**, or **super anti-fog goggles**?

Andrea is trying to help me get more **Quercetin**, but with global shipping bottlenecks, it might be a long shot.

In other countries, the message is murky.

Health officials contradict their own experts.

The **French government** claims COVID doesn't spread asymptomatically, while **Chinese epidemiologists** say otherwise—and warn of a continuum between droplets and full-blown aerosol.

I'm still urging my family and friends to act—*now*.

Social distance.

Avoid crowded, poorly ventilated areas.

Stop shaking hands—elbow bumps are in.

Cough or sneeze into your sleeve.

Wash your hands.

If you touch a public button or screen, sanitize.

Use gloves. Or a fancy Victorian scratching fork, I joke to a friend who can't stop touching her face.

We video chat with family, including baby **Ethan**.

At just a year old, he loves to say "Kai Kai" and claps along when we sing.

In South Korea, **850+ new cases** today.

They're shifting from containment to triage.

The sickest—those needing oxygen or ICU—will be treated first.

I read a study about a bus ride—an infected passenger spreads COVID-19 to someone **4.5 meters away**.

Even after the infected person gets off, the **viral cloud lingers**, infecting another who walks through it **30 minutes later**.

Avoid poorly ventilated spaces.

Say "co-morbidities" five times fast.

Mor-bidities, bidibies, dibites... yeah, I'm too tired.

I don't sleep enough, but I *do* study a lot.

Pathology at MIT, Pathoma.

I kind of get RNA viruses now. Transmission vectors.

Cabin fever fantasies swirl in my head—I imagine announcing the global spread like an old-timey **burlesque circus host**.

If your health department says they're ready, ask:

How many ICU beds are available today? Could you take on 1,000 new patients? 10,000?

Harvard epidemiologists say 40–70% of the world will get this virus in 2020.

If 20% need hospitals, for three weeks each, we're going to run out of runway fast.

Flatten the curve.

That's the goal.

Keep cases below hospital capacity. That's why quarantine matters. That's why China didn't overreact.

In Washington State, the virus's **RNA strand hasn't changed in six weeks**—which is good. Stable. But it's also been **circulating quietly** the whole time.

Vancouver's getting nervous. Shelves are going bare.

That's how it happens:

→ A case.

→ A few more.

→ A cluster.

→ A press conference.

→ Another cluster.

→ An outbreak.

Where are *you* on the curve?

I make a **peanut butter, honey, and banana taco** (banana unsliced, one slice of bread).

Time is short. Slicing a banana takes time.

After dinner, Shaolin and I relax.

Then—**Dungeons & Dragons on WeChat!**

Video chatting with six friends after **40 days of solitude** feels electric.

We laugh.

We roll dice.

We crack lame jokes.

It's glorious.

We take it slow. No splitting the party.

It's good to have a plan.

Meanwhile, health officials around the world give *mixed messages.*

In the Netherlands: *"Don't wear masks."*

Germany's CDC: *"Disinfectant is ineffective."*

Translation:

They didn't stock up early enough, and now they're gaslighting the public.

It's not right.

In California, two healthcare workers who followed droplet precautions tested **positive.**

Four days later, the CDC admits: *"Oops. That should've been aerosol."*

Now, **118 frontline workers are in quarantine.**

Off the board. Just when they're needed most.

If you fail to plan, you plan to fail.

Shaolin and I finish the night watching the rain and sipping tea.

I think I'll go soak some bean sprout seeds.

Tomorrow, I'll eat pancakes again.

And maybe, if I'm lucky, roll a nat 20.

March 3, 2020 – "Fail to Plan, Plan to Fail"
Tuesday – Day 39

COVID-19 is now in 70 countries, and more than 92,000 people are infected.

It's been raining heavily for days, and the air feels fresh. I pick up two packages: five pounds of bean sprout seeds and some organic apple cider vinegar. I'd asked Shaolin to order the vinegar, and she gave me a five-minute spiel about how my weird food is, well, weird—but she ordered it anyway.

Although I can't be sure from inside my little home, it seems like things in China are getting under control. On Monday, the World Health Organization (WHO) Director-General Tedros Adhanom Ghebreyesus said, "In the last 24 hours there were almost nine times more cases reported outside China than inside China." This shift shows quarantines work to reduce the rate of infection (R0).

I want to go to the mall soon—walk around, watch a movie, and have a coffee on the scenic cliffs of Hongyadong while watching the sun set over the river. I will do this again soon. Chongqing is waiting to be rediscovered.

We make delicious pancakes. I sip strong coffee while watching the news. At 2 p.m., I tutor eight-year-old Kim, who makes me laugh: "Hi fishy, nice to eat you!" Laughter is medicine.

Later, I scroll through online shops. Do I stock up on immunity boosters? A P100 respirator? Super anti-fog goggles? The "Hail Mary" Quercetin went to Canada. Andrea wants to help me get more, but shipping bottlenecks make everything uncertain.

In other countries, experts disagree. Some central health officials are saying one thing, while their top scientists paint a very different picture. The French government claims COVID-19 is not passed asymptomatically. In contrast, Chinese experts say the virus exists on a continuum—between respiratory droplets and aerosolized particles—and does occasionally spread asymptomatically.

I keep pushing my family and friends to act early, before a cluster becomes an outbreak. Practice social distancing. Avoid crowds, especially in poorly ventilated areas. Ditch handshakes; elbow bumps are in. Cough or sneeze into your sleeve. Avoid those who ignore basic hygiene. Don't touch buttons or shared screens—use gloves or a tissue, and wash or sanitize your hands often.

Nice pancakes for breakfast. Freshly grown sprouts for lunch.

One friend tells me she can't stop touching her face. I suggest a fancy, sanitized, Victorian-style scratching fork for her itchy nose.

Shaolin and I relax. We video chat with the family, including our grandson Ethan. Just a year old, Ethan loves saying my name—"Kai Kai"—and claps along when we sing.

With more than 850 new cases in South Korea today, the country is shifting from containment to triage. Hospitals will prioritize patients in most need of oxygen and life support. Other countries will soon have to follow suit.

A Chinese study documented aerosolized viral spread in a poorly ventilated bus. One person was infected from 4.5 meters away. The virus cloud lingered for 30 minutes after the infected person had left—infecting another who walked through it. Avoid poorly ventilated spaces. Say "co-morbidities" five times fast... mor-bidities-bibities-dibites... ugh. I get tongue-tied. I don't sleep enough, but I'm studying a lot. Pathology at MIT and Pathoma lectures are helping. I'm starting to understand RNA viruses and transmission models.

I have a crazy dream: narrating the global outbreak in a burlesque circus-style old-timey announcer voice. Cabin fever, maybe?

If you listen to your health department, they'll tell you they're ready. Ask two questions and watch the poker face crack: How many ICU beds are currently free? Could you handle 1,000 new patients? 10,000? 100,000?

Harvard epidemiologists estimate that 40 to 70 percent of the world may get COVID-19 this year. If even 20 percent of those cases need hospitalization, with each ICU stay lasting three weeks, we're in trouble. Don't panic, but be aware. This is why China's "draconian" quarantine wasn't an overreaction—it may have been a blueprint.

We're starting to talk about "flattening the curve"—keeping the number of cases low enough that hospitals can cope. This is vital.

A recent study traced the genetic makeup of cases in Washington State. Over six weeks, the RNA of the virus barely mutated. That means it's stable—which is good—but also that it's been spreading silently. Washington now appears to have community transmission. Vancouver is getting nervous. Empty shelves and growing paranoia say it all.

That's how it starts. A single case. A press conference. Another case. A few warnings. Then clusters. Then outbreaks. What level are you at?

I make myself a peanut butter, honey, and banana taco—just a slice of bread wrapped around a whole banana. Time is short, especially if you live it linearly. Cutting a banana takes time.

After dinner, we relax and then I host a Dungeons & Dragons session on WeChat. It's surreal video chatting with half a dozen friends after 40 days of solitude. We're excited to socialize with someone—*anyone*—who hasn't been locked in the house with us all month. We laugh, make lame jokes, roll dice. It's glorious. We try not to split the party and take it slow. It's good to have a plan.

Despite the calm, measured tones of many government briefings, doctors and scientists on social media are panicking. They're sounding alarms. They say there's incompetence and misinformation coming from the top. In the Netherlands, officials say handwashing is enough, and discourage masks or gloves. The German CDC claims disinfectant isn't effective. Why? Because they didn't stock supplies in advance—and now they don't want the public competing for what's left.

In the U.S., two healthcare workers in California tested positive after following droplet protocols with an infected patient. Four days later, they were told it had been an airborne case. One hundred and eighteen doctors and nurses were sent into quarantine—taken off the board when they're most needed.

If you fail to plan, you plan to fail.

March 4, 2020 – "The Usual"

Wednesday – Day 40
A real Italian quarantine.
I wake up, make some coffee, check my news stream. I post:
"Live stream coming up."
The connection's rocky, but I manage to chat with a few friends, and it's nice to feel a little less alone. One asks if it's still possible to order things from China. *Sure,* I say. "Factories are starting back up. Just pick airmail if you can—there might be a bottleneck of ships trying to leave Wuhan."
Then I notice something weird: **Facebook kicks me off whenever I say "Wuhan."** Like, three times in a row.
Censorship? Glitch? I don't know. It's surreal, especially since what I'm *trying* to say is things are **getting better here.** Still, it feels good to connect.
Shaolin's up by lunchtime, lounging peacefully.
We eat together.
I tune into **Super Tuesday**, and watch—aghast—as **Biden sweeps state after state.**
I'm not upset about Biden, not exactly. I'm upset Americans are voting **against their interests**—against healthcare, taxing the rich instead of the poor, and bold environmental policy, the **real existential threat** of our time.
Pandemics come and go. Climate collapse is forever.
Colleagues are freaking out over pay.
There's a legal notice that says foreign workers are guaranteed **70% of the minimum wage**, but that works out to about **RMB 1,250, or $250 Canadian a month.**
Good thing we saved for a rainy decade.
Last night's D&D was fun, but today I'm a little drained.

A low hum of **restlessness and creeping ennui** settles into my bones—until the **sun** cracks through the clouds and burns it away.
"Let's go outside," I say.
Shaolin hesitates, then says: "Let's walk to Ren Ren Le."
I'm not *in love* with the idea, but I say okay.
We suit up—masks, gloves, goggles—and head out.
Chongqing is slowly waking up.
The streets are still subdued, but **there's movement.**
Shops reopening.
People walking with purpose again, like they've got somewhere to be.
It feels like hope.
Everywhere we go: **temperature checks, forms, face masks.**
It's the new normal.
We bump into my friend **Patrick**, the vice principal of campus B.
He says school might start soon. That excites me—and terrifies me.
All it takes is **one sick student** and we're back at square one.
But today feels promising.
We pass a bubble tea shop. There's a **1.5 meter distance rule** and only **one customer allowed at a time.**
Progress.
We buy yogurt, processed cheese, and—miracle of miracles—**mushrooms are back.**
The supply chain lives!
Shaolin uses the touchscreen self-checkout, and I silently vow to buy a **stylus** just for that.
On the way back, I suggest crossing the street to avoid the **rank stench of a public toilet.** She doesn't want to—the left side's blocked off by police tape. Some building's been sealed, maybe a contamination case.
So I hold my breath and we hustle past the toilets.
She thinks I'm being paranoid.

But I think about the **microscopic molecules** that deliver smells to the brain—**nitrogen, oxygen, methane, hydrogen, aerosolized fecal matter.**

If I can smell it through my mask... doesn't that mean **it's getting through my mask?**

Smells are **inhaled**, after all.

We debate the science of it.

She shrugs. I grimace.

We keep walking.

You know, the usual.

When we get home, I listen to **RZA's Guided Explorations EP** again. It's still the best medicine I've found lately, right up there with vitamin D and filtered sunlight.

Later that evening, I shop for **new masks** online, browse P100 respirators, and toss around the idea of a **stylish hazmat suit**.

I fantasize about Hongyadong sunsets, bubble tea with no lines, and the day I can walk through the mall without a mask.

Maybe even sit down for a coffee like a normal person.

I'm not sure when that day will come.

But today gave me **hope.**

And for Day 40, that's a pretty solid gift.

March 5, 2020 – "Not Easy It Is to a Jedi Make"

Thursday – Day 41

Our Philips Air Purifier flashes its little red "Change Filter" warning.
No problem. I just bought a replacement.

I unplug the unit and begin dismantling it like a bored engineer on a long-haul starship. Q-tips, cotton balls, a dust devil vacuum, and some scrubby meditative focus. I wash the dust catcher and air out the HEPA filter on the balcony like a proud survivalist airing out his flag. A little bit of friction and flow, and soon it's all back together. Easy peasy. I'm officially the ship's engineer now.

We drink coffee and honey water and download hundreds of children's stories for our tutoring classes. I flip between the news and fairytales, shifting gears between **hope, imagination,** and **pandemic anxiety** like a multitasking bard.

When people ask me why I didn't leave China, I tell them the truth:
I stayed for **my wife,** and **my dogs,** and **my life here.**

Some foreigners left early on. Some mocked those of us who stayed.
Now, quietly, many whisper: *"It's safer here than where we ran to."*

Case in point: eight Chinese restaurant workers who fled Italy tested positive for COVID-19 upon return. China is focused now on **imported cases**—a reversal of the early narrative.

In economic news, governments are trying to **flood the zone with cash,** cutting interest rates and injecting billions to help small and medium businesses stay afloat. Some say it's early. Some say it's not nearly enough.

One medical study from Beijing and Shanghai confirms two distinct **strains of SARS-CoV-2:**

- The *L strain*: aggressive, contagious, but short-lived—burns bright and fast.

- The *S strain*: slower, steadier, sneakier—sticks around longer. As of today, we've got **97,000 infected worldwide**, with over **3,000 dead.**

It's been a few days since the dogs turned the floor into a bathroom. Maybe they've forgiven us. Or maybe they just forgot what they were mad about. Either way, the balcony routine works.

Big news today in medicine:

India restricts exports of 26 key Active Pharmaceutical Ingredients (APIs), used in antibiotics, antivirals, and more. They're not for treating COVID-19 directly, but vital in handling **opportunistic infections** that sneak in after a patient's been weakened.

Pair this with China's earlier supply disruptions, and **I hope people are stocked up** on essential meds.

I worry about my grandmother. Her monthly injections for macular degeneration keep her from going blind.

She can't miss a dose.

Seattle is still and silent. Like Daegu two weeks ago.

Italy's shutting down schools.

The U.K. is switching from daily to weekly reports—a subtle way to bury the bad news.

Friday releases are for coverups, not clarity.

I'm thankful for the Chinese apps that give me real-time maps of nearby cases.

Transparency matters. Public health must trump public relations.

Iran, overwhelmed, frees **54,000 prisoners**.

Europe teeters on the edge of its own China-style response.

We'll see who follows through.

Mask discourse in the West has hit a fever pitch.

U.S. officials say: *"Don't wear them."*

In Asia, we say: *"Wear one. Everyone wears one. Of course it helps."*

It may not be a Canada Goose parka, but a sweater is still better than nothing.

That's not panic—it's **math**.

I order gloves and a **heavy-duty mask** that doubles as a teaching aid.

Excited to test it out. Hope it fits this glorious dome of mine.

Meanwhile, in the U.S., a man in a **gas mask** is escorted off a domestic flight—not for being unsafe, but for freaking people out.

Seems like **feelings matter more than public health**.

That's how you end up with an outbreak.

Shaolin and I are doing well.

RZA's "Guided Explorations" is still my go-to therapy.

We prep for classes with calm hands and warm hearts.

Tonight's culinary mission: sweet potato fries.

Shaolin calls her niece Meito for guidance. We wash, boil, powder, freeze, and fry—*a multi-hour saga of tuber transformation.*

They're incredible.

There's chatter online:

- Some say the virus came from bats.
- Some say a biolab.
- Some say the U.S. planted it.
  The only official line? *We don't know yet.*
  All I know is: it's here.
  And it's still spreading.

Thermal imaging helmets are being tested by Chinese police to detect fevers in crowds. I love technology. I just wish we had more of it sooner.

I go off the rails a little tonight and mix **hummus and salsa into my pasta.**

I also make a **wasabi mayo** for our fries.

It's wild.

But it works.

Our home isn't big or fancy.

But it's optimized. It's **clean**. It's ours.

It's our spaceship.

And on this ship, Shaolin is **mission control**.

Sometimes hard on me. Always pushing me to grow.

Think Yoda, but stylish, stubborn, and with better dance moves.

I'm trying to be a good Jedi.

After all, as any apprentice knows:

"Hard it is...

to a Jedi make."

March 6, 2020 – "It Takes a Village to Catch a Virus"

Friday – Day 42

Today marks Canada's first untraceable COVID-19 case. No travel history. No contact link. Just a person, infected — and we don't know how. **Community spread has officially begun.**

Thirteen new cases today. Four provinces affected: B.C., Ontario, Québec, and Alberta. In Vancouver, two schools close after a presumptive case. In Kitchener-Waterloo, the virus hits and schools follow suit. Toronto adds another case — this one linked to a trip to Las Vegas. The USA is now exporting infections. Travel history is becoming irrelevant.

"We're not in containment anymore," says Dr. Colin Lee, a Canadian infectious diseases specialist. "This virus can now spread invisibly."

Dr. Lee calls this a *sentinel event* — a signal that the game has changed. We're shifting to mitigation mode now: school closures, mass gathering bans, emergency planning.

"We still have time," he says. "Look at your household. Get your supplies ready."

It's surreal to hear such words from mainstream experts, when just weeks ago they were still downplaying it. "It's not as bad as the flu," they said.

I check in with Jay. He's got the high-end masks I helped source, but things didn't go as planned. He had organized a private charter out, but his wife didn't want to go to the U.S. She chose her family's countryside home in China instead — it felt safer. Now Jay's stuck in solo quarantine in Malaysia, planning to head to LA to ready the house for their arrival.

As stores run out of everything from **toilet paper to frozen pizza**, I hope supply chains hold. China kept fresh produce flowing during lockdown. Can Canada do the same?

Even my dad — an ardent Trudeau supporter — is annoyed. Trudeau shrugs off travel restrictions:

"There's a lot of misinformation out there... knee-jerk reactions... We're going to stay focused on doing the things that actually matter."

I wonder if he's waiting for it to get *really* bad before doing anything. It wouldn't surprise me.

In the U.S., the situation is worse. They turned down the WHO's working test. They wanted to create their own universal test for all coronaviruses — ambitious, but flawed. It failed. Now, weeks later, they still have **no accurate measure of their infection rate.**

"We have no idea where this thing is," said an American health expert on TV. "This thing is *everywhere.*"

We teach a class tonight, so I try to chase the daylight. I skim wiki lore for a game I'm considering writing for, to weave my story arc into its threads. As long as I've got **avocado toast and strong coffee**, I can quarantine with dignity.

Shaolin is flipping through TikTok while I tinker with *Amos*. I don't get as much done as I want. I mean to. But it's one of *those* days.

Texas handwashing instructions:

"Wash like you just diced jalapeños for nachos and now you've got to take your contacts out."

Financial advice? Sure.

Invest in protective gear, streaming services, and online shopping. But real wealth? A **cottage**, with a **well**. That's the dream.

The sun breaks through and I ask Shaolin to come outside. She declines. Then, 20 minutes later, she gets suited up. "I'm going out," she says. So I follow.

We sit in our chairs under a tree. I read *Agency* by William Gibson. It's so good. I think I'm reading it slowly because I feel like I'm *in* a Gibson novel right now — mask on, synthwave soundtrack, cyberpunk reality. I keep one eye on the news. My day blends Chinese, Korean, Italian, Singaporean, Indian, French, British, and American briefings. WHO,

CDC. Dr. Chris Martenson. Dr. John Campbell. Ten hours a day of
multisource information synthesis. Luckily, I'm a multitasker.

The sun peels back my anxiety. I strip down to a muscle shirt, my jacket
slung over the rail.

Shaolin sees me.

"Put your clothes on. It's not summer."

I laugh. "Feels like it."

She frowns. "Put your mask on. You'll scare the people."

"What people?"

She points to the apartment across the way.

"There's no one in the windows," I say.

"They're watching," she says.

I chuckle and put the mask back on.

In Canada, people are getting kicked out of stores for **wearing** masks.
Here, I get scolded for **taking mine off for five minutes in the sun.**
The irony is almost too rich to swallow.

Shaolin bakes a cake. I distract her briefly with a fridge-smell test, long
enough to jot down notes. We crack eggs, stir in flour and bean powder.
After class, we try it — like cheesecake without the cheese. It's *excellent.*
Dinner is salad and pasta. My real crisis? **Four ripe avocados.** Tomorrow, I go wild.

A bright note: after some emails with Airinum, the Swedish upscale
mask company, they're replacing my broken filter with **ten new ones**
— a miracle, considering they're sold out globally until July. Each filter
lasts 100 hours. With my 1.5 filters and current 30-minute/week exposure, I now have 1,150 hours of coverage — **44 years of protection** at
my pace.

"Sounds crazy," I tell Shaolin.

"Not crazier than catching it," she says.

Shaolin's feeling better. No sore throat tonight. Her shoulder's improving.

I debate the video game writing job again.

Pros: great money, creative work, no admin nonsense.

Cons: I'd lose my free housing and four-month vacation.

More importantly: I'd be writing **someone else's story**, not my own.

Shaolin says my hair's turning white in the back.

"Really?" I turn in the mirror.

She smirks. She just likes watching me spin.

Outside the window, the streets are empty again.

And still, my friends are **traveling**.

Conferences. Cruises. Seattle for fun.

They don't get it yet.

"If they just stayed home..."

"They wouldn't get sick."

It really does take a village...

to catch a virus.

March 7, 2020 – "The New Normal"

Saturday – Day 43

I wake up tired but excited for the day. I scroll through the socials, dial in the news, and fire up some strong black coffee.

102,000 COVID-19 cases.

Did the first 100,000 get a balloon drop and a video chat with Dr. Tedros?

The virus has now reached over 80 countries. I scan the data. About 15% of cases are severe or critical — down from 20%. That's something. The case fatality rate is anyone's guess: WHO says 3.4%, the U.S. says 0.5%. Italy's numbers are high. Could they be dealing with the more aggressive L strain? South Korea, by contrast, has things relatively under control — every new case linked to a known cluster.

At 10 a.m., I slip into the first of three two-hour classes. All's well until I dump an enormous mug of coffee all over Shaolin's prized bedroom ornaments. Her eyes narrow. I clean it up, muttering about how coffee is also a good surface disinfectant. Turns out I've had enough caffeine.

Dr. John Campbell says he expects to get infected in the next few weeks. It breaks my heart. His models predict 75% of the global population may get this virus. Inescapable.

I spend hours chatting with friends — some panicked, some oblivious. I share my protocols. "Outside clothes vs. inside clothes" trips them up. They have kids. I get it. Still, I prefer rigorous discipline. I don't just hope to stay safe — **I prepare to stay safe**. This is the way.

Dr. Campbell praises China and Singapore's response: screening, social distancing, self-quarantine. He's far more critical of the CDC's botched testing rollout in the U.S. Countries like South Africa worry me — immunosuppressed populations may see much higher mortality rates.

A new study shows that wearing a mask around respiratory-infected individuals is **80% more effective** than nothing. We've known this in Asia for decades. North America is still debating it.

Toronto is now disinfecting subways. A case exported from LA had ridden several lines before isolation.

Japan: "We'll never live down the Diamond Princess disaster."

USA: "Hold my beer."

Now it's the *Grand* Princess. Another cruise, another outbreak. Over 20 infected crew. Here we go again.

I gear up and head out for a package, stylus in pocket. No need to touch screens anymore.

I feel calm. Controlled. Prepared.

RZA's meditations in my ear. This is how I warrior.

Lunch is noodles, with chickpeas and top-shelf hummus. If you haven't had a chickpea sing on your tongue, you haven't lived.

I ping my dad. Amazon packages arriving next week. I remind him: **Disinfect.** Use gloves. Open outside. Leave boxes to "cool." Wipe down contents. Wash hands. Simple. One Amazon worker in Seattle's already in quarantine. That's a *prime* delivery.

I set up cushions in the sunbeams for the dogs. Today is 20°C and blindingly beautiful. I remix my protective gear for the heat.

Summer mode activated.

We go upstairs to the garage — now renamed **"The Gym."**

The whole rooftop shimmers. Beach vibes.

Shaolin and I sit. She stretches. I lean against the rail like James Dean. I take off my jacket, strip down to my undershirt. I even take off my mask.

Reckless?

No.

Intentional. Sun-starved.

I read *Agency* by Gibson — cyberpunk at its finest, in a world that now mirrors his fiction. I smile. I feel good.

A breeze tickles my shoulder. I jump. It's just my hat, swinging on a tree branch.

We're cool.

After reading, I stretch, pace, and run through my kicks.

Bruce Lee once said:

"I don't fear the man who has practiced 1,000 kicks once. I fear the man who has practiced one kick 1,000 times."

That's the goal.

Front kicks. Side kicks. Roundhouse. Hook. Back kicks.

My family trained with Taekwondo Grandmaster Tae E. Lee. Thirty years later, I still carry his discipline.

Back home, I wash my hands — singing *Bohemian Rhapsody*. Not just the chorus. The whole thing.

20 seconds? That's nothing. I'm in it for the full performance.

In the shower, I close my eyes and imagine I'm blind. Practicing presence. Empathy.

Afterward, I swear I see the beginnings of a tan. A little sun. A little shine.

I put on my talismans:

- **Black obsidian**, blessed by a Mayan medicine man
- **Horseshoe with blue topaz**
  A little magic never hurts.

Chongqing update:

- **576 total cases**
- **50 hospitalized**
- **3 severe, 1 critical**
- **6 deaths**
- **520 recovered**
- **Zero new cases for 11 days**

We're holding the line.

My protocols are solid. So are my convictions.

Mask debate.

I've heard it all. "They don't work." "You'll touch your face more." "They're for doctors."

But here's the truth:

Yes, a 0.1-micron virus particle can fit through a 0.3-micron filter.

But that's not how viruses travel.

They hitch rides on droplets — coughs, sneezes, breath. Masks stop most of those.

Less viral load = fewer complications.

Less virus = better outcomes.

Wearing a mask is like putting a bouncer at the entrance to your lungs. Will it stop everything? No. But it stops *a lot*. And that may be the difference between mild and severe illness.

I talk with Dr. Victor Wood, a retired ER physician. He confirms everything. He adds:

"Masks also remind you not to touch your face."

Touching your face with contaminated hands is a huge risk. Masks stop that too.

The rules:

1. Put the mask on. Make it snug.
2. Don't touch it. Don't fidget.
3. Don't take it off until you're home.
4. Use the ear straps only.
5. Dispose of it or isolate it for 10 days in sunlight.
6. Don't think it makes you invincible. Stay cautious. Stay distant.

Things in Chongqing are moving. Factories are restarting. People are returning to work. I'm wary of a second wave. The classroom is my personal cruise ship — I'll remain vigilant.

We go out for bubble tea. Everyone's masked, except one delivery guy cruising by with his nose hanging out like a rebellious child. I recoil. He's a half-masked terrorist.

Tuck that honker in, buddy.

Next week we've got errands. Bank. Condo. Tax office.

I'm not ready.

RZA help me.

I am the sun, surrounded by planets, space junk, and chaos.

I am the sun.

We flip through our digital photo frame — 2,000 memories of vacations, family, pre-pandemic joy. It's comforting. We've been to a dozen countries, danced on many continents. And we will again.

But not yet.

We make pancakes for dinner.

Because we're adults.

Because we can.

Shaolin wants mindfulness. I want to multitask.

We compromise. I stir slowly and pretend I'm a man who's traded his riches to return to this moment — strong, healthy, full of life — just to kiss his wife.

That's what love looks like during the end times.

This is maintenance. This is devotion. This is survival.

From vitamins to vinegar, clean sheets to moisturized skin, *everything* I do is in service of a summer flight to Canada.

Until then, I'm solid.

I'm the sun.

And I thrive in chaos.

March 8, 2020 – "A Life Half Lived"

Sunday – Day 44
International Women's Day

Today is International Women's Day. I make Shaolin a whipped milk sugar latte and toast up our homemade egg cake for breakfast. She loves it and takes a few pictures for her peeps. I drink strong black coffee and eat some boiled eggs with guacamole toast. Shaolin looks lovely today. We are both content and peaceful—a nice turn for us.

We've had almost 50 days inside since I finished teaching my fall semester. We didn't get a nice winter vacation this year, but it's been relaxing in its own way—once you remove the anxiety, fear, unknown, panic, prepping, and hoarding. You never know when you're going to wake up and everything will change, so enjoy every single minute.

My favorite astrophysicist, Neil deGrasse Tyson, was on Colbert. He was great, as always. "We are in the midst of a massive experiment," Neil said, speaking about the global response to COVID-19. "Will we listen to scientists and take their precautions and instructions to heart?" We will—either out of respect or fear for the consequences if we don't. "But we shouldn't be too afraid to live. A life lived in fear is a life half-lived," Neil added. We can be aware, and we can be alert, but we should not let fear consume us.

In the pursuit of a virtuous life, I look back to the ancient sages of the old world. *"If a person gave away your body to some passerby, you'd be furious. Yet you hand over your mind to anyone who comes along, so they may abuse you, leaving it disturbed and troubled—have you no shame in that?"* – Epictetus, *Enchiridion.* We protect our possessions and our money, yet we give away our time and our minds to anyone who presses upon us. It's essential for a creative person, or an entrepreneur, to learn how to say no. Vampires were always metaphors for viruses—you have to invite them in.

When the virus comes to my door, I'm not opening it. I won't allow myself to think about it all day either. I'm still interested in learning, but there's an acceptance growing within me. I must learn to balance my energies to make it through the long haul unscathed.

There are 108,000 cases of COVID-19 in the world today. Many people are alarmed over Italy, where 18 million people are now under quarantine. They call it a "soft quarantine," because Milan's airport is still partially open. There are too many cases in more than 80 countries now to list them all.

The CDC is telling us to hide our grandparents in the attic to protect them.

The colossal debate about masks keeps coming up—Asian countries versus Western ones. Asia believes in universal masking to cut the spread on public transportation. In the West, it's "masks don't work unless you're a doctor." I can't understand it. I've engaged with Donald Trump, the U.S. Surgeon General, and even Elon Musk this week trying to understand what they're thinking. Still waiting on a satisfying answer.

At least here in Chongqing, our procedures are working well. It's been almost two weeks since our city has found a new infected person.

Before we teach our 1–3 p.m. class, I play a game of hockey online with my Pops. He whoops me, which is a nice change for him—and I'm a good sport about it.

I found my hefty three-pronged MacBook adapter, and I feel more ready to handle a pandemic now. Pandemics require a hefty approach.

Yesterday, the Xinjia Express Hotel collapsed in Quanzhou, Fujian Province. The CDC was using it to quarantine people who had possibly been exposed to the virus. More than 10 are dead, and 25 are still missing as the rescue effort continues.

Our first class is easy and fun. We have a four-year-old student named Elsa, and she's a joy to teach.

The second class is more challenging. I have 11-year-olds who struggle with basic words and forget everything by next week. Rather than be tormented by the frustration, I imagine they are just lackluster demons, doing a poor job of torturing me in some sad little circle of hell. I will emerge unscathed.

After a 10-minute break, I switch to the textbook and things go smoother. I realize I need to keep it simple. It turns out I'm the idiot—it's easy to criticize when things aren't going well, but it takes a real leader to step up and fix them. That's a lesson that's universally applicable.

Last night we watched *All the Bright Places* on Netflix. The film was well cast, well written, based on a novel. It's terrific, although a bit sad. I suppose we can't have happy endings anymore. Somebody decided the kids won't believe in them.

After work, we order pizzas and Shaolin makes dumplings. I toast the pizza, just to be safe.

Tonight, we're going to relax—read a bit and watch some movies. Tomorrow's a big day. Lots of errands, far from home.

March 9, 2020 – "Sky Walker"

Monday – Day 45

I wake up in a deliciously liminal state, before awareness floods in. Home. Loved. Safe. Normal. Then, like a meteorite crashing into earth, everything hits me—and I take solace in strong coffee and a delicious brunch.

The tension between my cutting sarcasm for the farce of my existence and boundless empathy for the untold human tragedy I witness is the blade of a knife I dance on. Would you like some tea?

I summon RZA daily now. His smooth cadence and guided experience ground me. Mindfulness meditation WuTang HipHop—2020's surprise genre mashup I never knew I needed, and would probably go insane without.

I tidy up, drink more coffee, and teach a class about daredevils. My colleague Michael needs another file from his laptop. The fact that I can just dial it up now, when that used to bother me a lot, means I'm coping better.

I do some laundry. I wash my space suit travel onesie.

The Grand "Disaster" Princess II is evacuating the sick first in SF—so they can touch everything on the way out before the possibly healthy leave. Unless they have separate exits, this is a disaster. I worry about them. I worry about all of us.

My worst habit? Wiping my hands on my pants. Chip crumbs, sticky fruit hands, whatever. When I'm making pizza or bread, it's a mess. So they're hanging to dry, and tomorrow I'll try again.

I'm watching my bananas turn brown, but I've got a plan.

In my life, everything happens quickly. Even ice caps melt fast these days.

My cousin Andrea messages me about the first case in Minneapolis. She wants to double-check her prep list. I'm happy the family is prepping.

Half the U.S. states now have at least one of the three Cs: cases, clusters, or community spread.

Cousin Scott and my friend Sebastian are both conservative-leaning but smart. We touch base. I'm happy to hear from them. I'm becoming less polarized. I may be locked inside a literal tin can, but I'm out of my echo chamber.

In Chongqing, a live broadcast of medical professionals via online meeting software makes me proud of how we're handling things here.

Meanwhile, flights out of Northern Italy are flying into the U.K. with no screening. Public Health England says to self-isolate for 14 days. How many are asymptomatic and at work now? Why does everyone try to flee quarantine? Don't they know that's how it spreads?

"If you hear of an outbreak of plague in a land, do not enter it; but if the plague breaks out in a place while you are in it, do not leave that place." – Muhammad, Prophet.

Do the right thing.

80% of cases: smooth. 15%: rough. 5%: life and death. Which city can put 5% of its people into the ICU? Models predict a collapse of the U.S. healthcare system by May. What are they going to do? What are any of us going to do?

SXSW: canceled.

Ultra: canceled.

Other festivals: "Nope."

Leslie and I are canceling Burning Man. I tell Maid Marion, maybe we'll all burn two meters apart this year, masks on. Entry process: 14 days, everyone stays in their car.

Dr. Bonnie Henry broke down and cried on TV. Some say "what empathy." Others say "we're doomed." Hard to know what to think.

We pack for our spacewalk. My anxiety is palpable.

I've been safe for almost 50 days because I've stayed home. Now I'm going out, expanding my universe from the comfortable, tidy, life-nurturing spaceship that has been home through this crisis—into the unknown.

A bee buzzes. The city hums.

I summon my teacher. The RZA is beside me. With him, I will kill the noise.

Welcome to the exploration.

Shaolin calls a Didi. We walk to the gate.

"So far, so good," I tell myself.

Today, we face a familiar adversary: distraction.

As we get to the gate, I realize—this is usually the end of the journey. Not today. Today it's just the beginning.

We get a tracking paper and leave the compound. I take a short video of kids playing and people just... living.

Five minutes later, our car arrives. Shaolin tries to sit in front—motion sickness—but the driver says no. A plastic sheet separates him from the back seat. There's a bottle of cleaner. So far, so good.

RZA whispers: "Before we take off, ground yourself."

I breathe slow. In through my nose. Out through my nose. Garlic breath contained.

We drive across Chongqing from Jiulongpo to Yubei.

It's not a ghost town. People are doing their thing. But everything has space—on the road, on the sidewalk, between everyone.

At the compound, a guard doesn't recognize me. Shaolin explains I'm her husband. I sign a paper. We take our temps.

We meet the rental agent. Foot tap greeting. We go up. The elevator's button panel is covered in chopsticks—COVID-style.

The condo looks good. Bit messy, not dirty.

We clean. We laugh. We relax.

The woman renting it arrives tomorrow. From Harbin. Canadian-style winters, six-foot-tall women, great dumplings. She'll quarantine here 14 days before work.

Later, another Didi back. No divider in this one, but it's coming. Everyone's adapting.

On the way home, we grab snacks, drinks, two packages: avocados and some pot stabilizers.

We cook hot pot and pizza. We eat. We rest.

I edit some news. Talk with Rhett Morita—he's got a samurai Netflix doc coming. He's prepping too. He liked my stylus trick.

A friend in Seattle got COVID. She went to a party. Everyone seemed fine. A few days later, half had flu symptoms. She tested positive. Got through it with Sudafed and a Netti pot.

I work on a video project until 1 a.m. Then, my blog. It's late, but I love the rush.

I miss all-nighters, making tracks. Shaolin thinks I'm nuts—but she's cool with it.

RZA: "You are the sun. Everything else is orbiting."

I breathe.

Thanks, RZA. You got me through today.

See you tomorrow.

All right.

March 10, 2020 – "Drive-by Sneezing"

Tuesday – Day 46

I wake up exhausted from the incredibly full day I packed into Monday. I've missed the blending of cycles, where work bleeds into sleep and appetite—the drive. After some strong coffee, the cobwebs are wiped away. There's nothing on my agenda until "D&D: 8 p.m.," and that feels great.

I check in with my editor; she takes a look at the nearly 3,000-word mess that was Monday. We tidy up a bit and then I dive into editing my *Drive-By Jorah Kai* video, laying in tracks, ironing out titles, and rendering footage.

Shaolin gets a call from a community leader from our condo building. The woman sounds alarmed after seeing a foreign name on the entrance sheet yesterday.

"Who is this foreigner? Is he new to Chongqing? Why is he inside our building? Did he rent the condo?" she fires off rapidly.

"No," Shaolin says. "He's the owner. It's his house. He's my husband."

Unbelievable. But they're tightening up. There's concern about foreigners coming back from hot zones.

I drink so much coffee I feel like a meat straw. Should I be eating more? Later, I go out to get a package and hear a sharp intake of breath. I turn and see a middle-aged man pull down his mask and sneeze, twice, into open air. The sound ricochets between buildings. Everyone freezes—deer in headlights of this bioterrorist's breach of public decency. A COVIDIOT.

New studies show COVID-19 can travel up to 4.5 meters in some conditions.

I want to tell him off, but my brain runs the math: angles, wind vectors, Pythagorean theorem in a matrix-style slowdown. I'm on the edge of his sneeze zone, so I jog away, leaving the others to fate.

Back home, I decontaminate, but I feel a sneeze coming near Shaolin. I run to the bathroom and sneeze into my shoulder five times. Then I scrub myself with hot water and soap until I finish *Bohemian Rhapsody*. Later, I help proofread a Coronavirus project for my editor—second sets of eyes are always helpful.

Introverts across the world rejoice as gatherings are canceled. 2020 is canceled.

We walk to the bank for Shaolin's errand. It's a nice, sunny 25°C stroll. A few days ago, this would've felt huge, but I'm more relaxed now. Still cautious.

At the bank, Shaolin signs in while I'm waved away. They don't want to deal with a foreigner. I understand. After that phone call, everyone's on edge. I wait outside, dance-walk to our Root Sellers grooves. Some police give me weird looks. An hour later, she comes out. "They were alarmed," she says. "Even with all your gear, they somehow knew you weren't Chinese."

"How?" I ask. "I was wearing goggles and two masks."

"They asked if you went back to your country. I said, 'No, we've been here the whole time. Forty-five days.' That calmed them down."

Back at the gate, drenched in sweat, I panic. If I fail the temperature check, they could drag me away before I explain it's just heatstroke. I peel off layers, wipe down, cool off. Luckily, I pass.

We pick up a package and go home. After changing and scrubbing up, I render the video again, then head up to "the Gym"—our rooftop garage. The sun's out, I'm smiling. Life is good. But my face has "goggle burn."

We eat fresh greens and rice for dinner. Later, I catch Shaolin on the bed in her outdoor clothes, video-calling her sister. I freak out, yell, and sanitize the bed. She's letting her guard down after 50 days—I'm not.

Then, an internal conversation:

My agent/PR persona says, "Kai, the whole 'I'm in quarantine, look at me' thing was cute for a month, but it's getting old. Italy's in quarantine now—they're the hot new show. What are you bringing to Season Two?"

"Survival? Positivity? Writing?" I reply.

"Boring," he scoffs. "Let's give you COVID-19."

"What?!"

"Yeah, just a little! It'll spike the drama. Book deal, boom!"

"No."

"How about licking a subway pole?"

"No."

"Cut holes in your mask?"

"No."

"You're no fun," he sighs. I go help Shaolin with dinner.

All of Italy is now quarantined. Sixty million people. It was obvious. Let people flee the hot zone, and the virus follows.

"The right decision is to stay home. Our future is in our hands."

—Giuseppe Conte, Italian PM

An Italian prison riots. The NYSE halts after a 2,000-point crash—Black Monday 2020. The pin that popped the bubble is COVID-19. The golden needle we weren't looking for.

In Chongqing, cases continue to drop. Of 576, only 50 remain hospitalized, most discharged or recovered. We're almost back to work—but guarding our city's gates carefully.

We'll make soup, banana bread, and pineapple pie tomorrow. Big plans for a quiet day.

Shaolin does salsa online. I snap a photo—she's glowing.

At 8 p.m., I suit up as Fis the Fierce in our D&D campaign. Afro Samurai. A fierce joy.

After the game, I finally upload the *Drive-By Jorah Kai* video to YouTube. It gets banned. I'll try again tomorrow.

At midnight, I relax with Shaolin. Life's weird. But I'm grateful.

March 11, 2020 – "Decoding the I'm Possible"

Wednesday – Day 47

I'm going to let you know three impossible things before lunch: We never really die, we just change, and time is only an idea—it couldn't hold a bucket of muck.

Today, I finally slept decently and am striking a balance between trying to save the world and enjoying a hearty brunch. Have I mentioned I'm a big fan of coffee? It's worth saying again.

I watch some Pathoma—I'm still trying to demystify the human body and how we get sick. I take a beautiful hardcover edition of *Dante's Inferno* off the shelf and spend some time relaxing in Hell.

Italian soccer player Daniele Rugani announced that he has the coronavirus. After Daniele's positive test, the Series A league, where he played for powerhouse Juventus, suspended the season. "I'm fine, I've been pretty good. I hope it will serve to raise awareness," he told the team's football channel after a few days of self-isolation. "I've practically finished all of Netflix!"

Trump has said he's been tested and does not have COVID-19.

There's a video going around of Italians singing from their balconies during quarantine. It makes the hair on my arms stand up and my skin tingle. Two months ago, I saw the same thing in Wuhan. People are the same all over, and music is medicine.

I send out a couple of letters—one to strengthen government policy based on the proven stringent methods used in China, Singapore, and South Korea. It really is a gold standard. If enough people in Canada and the West take heed and push their leaders for more control, it could make a big difference.

I teach Lil' Kim for an hour. It's hilarious and fun.

The sink is leaking, but then—I fix it.

Today, the Level 1 emergency status in Chongqing is reduced to Level 2 because we haven't had a new COVID-19 infection in 15 days. We're going to Shaolin's parents' house tomorrow! That's big news for me.

We're trying to register me in the "Antivirus App" that will grade me green, orange, or red depending on where I've been and how long I've stayed in Chongqing. Since I've been at home for 45 days, hopefully I get green. I sort of wanted to avoid being registered at all and live off the grid like some kind of Luddite hermit. Still, if I want to see family, it's the only way in.

A couple of packages arrive—now that's no sweat for me. One is my full-face respirator. I originally bought it in case I had to teach. Now I guess I'll keep it for my trip back to Canada. Good to have. It took so long to arrive I almost requested a refund, but I held out. Leap of faith. I play with it, install the filters, and try it on. It fits well. This thing means business. The hard outer shell will be easy to disinfect, and I can tell it works because when I press my clean hands against the filters, I can't get any air at all. A perfect seal. With two antivirus filters, it'll last for years with minimal use.

I send out some letters—one to policymakers about how to equitably deliver capital to citizens about to be out of work, and another to people who want to avoid infection: a 10-point plan. We share a petition. Hours later, it has 50,000 signatures and growing strong. A mayor in Australia and a member of parliament in Canada reach out to follow up. Later I chat with a friend working with Bernie Sanders to save America. A good day for meaningful outreach.

Then, I make a loaf of really tight banana bread. Super grandma vibes. I'm going to call her and tell her all about it.

I think my video's done. It won't be the last one, but I'm so tired, I won't make another for a while.

There's an app that color-codes you by risk level, triangulating your GPS over the past few weeks. Not having a Chinese ID makes for frustration, but after a while, Shaolin and I figure it out—input my phone number, passport, and bypass the ID section. It churns for a moment, peering into my movements, and quickly judges me a serious hermetic recluse.

Green.

I can travel freely in Chongqing now.

March 12, 2020 – "In Chongqing, We Trust"

Thursday – Day 48
Part I: The Water Is Rising

My heart bleeds for Italy. Just seven months ago, we waltzed through dreamy piazzas as if we had all the time in the world. Now, Italy is under siege by an invisible alien force, collapsing from every side as COVID-19 overwhelms its healthcare system. Italian Medical Chief Roberto Stella has died from the virus at age 67. RIP.

On March 12, news broke that Mikel Arteta, manager of Britain's famed Arsenal Football Club, had tested positive. He was reportedly doing well. "The truth is that we didn't expect this virus to reach the club, the locker room," said player Lucas Torreira. Arteta became the first Premier League figure to be diagnosed.

Some criticize China for moving too slowly at first. Fair enough—we had no idea what we were up against. But what about the rest of the world, watching as China locked down hundreds of millions in a Herculean effort to contain the outbreak? Did they act swiftly? Many dismissed it. "Oh, that's China." Italy didn't worry. Their youth laughed it off and partied in public until the "viral bomb," as their top doctors call it, detonated—leaving them to make impossible decisions. Now, America and Canada have the benefit of hindsight. The question is: are they doing the right thing, or are they waiting for the flood to hit?

In northern Italy, hospitals are stretched to 200% capacity. Patients sit in chairs with oxygen tanks because there are no beds. It's a grim echo of Wuhan's earliest days—maybe worse. Patients over 65 are no longer assessed. Let that sink in: in Lombardy, once the domain of the powerful and wealthy, elders now die gasping for air, not because they couldn't be saved, but because someone younger might be. It's triage at its most brutal.

Doctors and nurses are getting sick but keep treating patients until they collapse. Only 10% of non-COVID hospitals are screening health-care workers. Forget strokes, car accidents, or routine surgeries—those aren't being treated. This isn't just a wave. It's a tsunami.

And still, some call me an alarmist.

But exponential growth doesn't care about your opinion. Here's a thought experiment, adapted from Professor Albert Bartlett at the University of Colorado: Imagine you're in the highest nosebleed seat at Yankee Stadium, and a single drop of water lands on the field. Then, every second, the number of drops doubles. Two, four, eight, sixteen. When would you notice?

Less than one hour later, the stadium would be flooded. You'd sit there oblivious for 54 minutes, then have just six minutes left—along with 54,000 other people—to get out before drowning. That's the terrifying truth of exponential growth. As Bartlett put it: *"The greatest shortcoming of the human race is our inability to understand the exponential function."*

"Wait and see" means "wait to die."

If I had the power, I'd shut down all nonessential travel now. Everyone entering a city should do a mandatory 14-day quarantine, like we do in Chongqing. Suspend tourism. Cancel gatherings. Yes, it sounds extreme—until you realize the alternative is mass death. Everything I've done has felt like too much in the moment, but I know already it will seem like too little in hindsight. Exponential growth can be our ally, but only if we move fast.

Canada's young, charismatic Prime Minister, Justin Trudeau, didn't want to make a "knee-jerk" reaction by banning travel or canceling events. Today, he and his wife Sophie are in self-isolation. Hours later, Sophie tested positive for COVID-19.

A knee-jerk reaction is exactly what you do when you've been knocked to your knees.

Ontario has announced that schools will stay closed for two weeks after March break, yet Premier Doug Ford cheerily encouraged people to "have fun" on their vacation. Will Trudeau channel his father's iron spine and invoke emergency measures to lock the country down? If he stood before us and said, "Just watch me," I'd respect him for it.

A century ago, the 1918 Spanish Flu pandemic offered two case studies. Philadelphia let the virus run rampant. Bodies piled up in homes. The morgues overflowed. In contrast, St. Louis implemented social distancing and flattened the curve, giving hospitals a fighting chance. With COVID-19, about 15% of people develop severe symptoms, and 5% need intensive care. The WHO's current fatality estimate is 3.4%. Without flattening the curve, up to 15% of critical patients could die—not because of the virus itself, but because we failed to save them in time.

Think of the difference between a five-minute ambulance response and a one-hour delay.

Flattening the curve is no longer optional. It's a civic duty. It's a moral obligation.

Hospitals don't bounce back overnight. Critical care patients can occupy ICU beds for weeks. That's why a slow, long curve is far better than a short, sharp spike.

In the face of limp-wristed leadership, people in the West have launched grassroots efforts like Stayhome and staythefuckhome.com, helping citizens understand how staying inside—right now—saves lives.

St. Louis or Philadelphia. Take your pick.

What can you do? Work from home. Cut your spending. Fill your pantry. Stay inside.

Become a Pajama Hero. Because pajamas don't just save your sanity—they might save your neighbors' lives.

Part II: The Storm Hits Home
Shaolin is getting ready to visit her parents. We're both exhausted.

My video is finally done—35 minutes of my life in Chongqing, a fortified city that has taken extraordinary steps to stay ahead of the curve. Quarantine for all new arrivals. Temperature checks at public areas and residential gates. Contact tracing. Testing. I scrub every shaky camera angle from my phone into this strange little documentary and set it to music from my own band, The Root Sellers. I try to upload it.

YouTube immediately blocks it.

Apparently, I'm not allowed to use my own music until their legal AI overlord finishes grinding my soul into metadata. I reach out to friends—label owners—to get the video whitelisted. While I wait, I feel raw. Shaky. Sleep-deprived.

I go to bed late. Shaolin is grumpy. She's right, of course. It's not healthy.

I get up at 4 a.m. and try again. I sleep for a few more minutes, then rise and try once more. I can't rest—not until I've done everything in my power to share this message. For the first time in 25 years in the music business, this mixtape might actually save lives.

Some people have acted on my warnings in recent weeks. They've thanked me. I haven't answered all the messages—I'm overwhelmed—but their gratitude keeps me going.

I haven't slept, but I'm strong.

"Last night a DJ saved my life" has finally come true.

In a strange twist of fate, publishers are reaching out. Some want to turn this blog into a book. A movie. Usually, I'd be thrilled. But today I'm just... resolute. I won't revel in the chaos—but I'll lean into it. The obstacle is the way.

I'm up past 5 a.m. Yesterday. This morning. The days are blurring as I fight YouTube's copyright bot, pleading for permission to share something that might help someone, anyone. I feel like a desperate fool, hoping my tiny drop of good information can hydrate a parched, disbelieving Western world.

Dov at Muti Music. Myagi from Pop and Lock. Westway. They're helping, but it's slow. And people are dying.

Meanwhile, in the U.S., Rudy Gobert becomes patient zero for the
American sports world. A few hours before tipoff, the NBA cancels the
Jazz vs. Thunder game. Gobert had joked about the virus the day before—licked his hands, touched all the mics. Now the entire Utah Jazz
team is quarantined. The NBA suspends its season. The ripple begins.

John Aquaviva, an old friend and house music icon, goes public with
his own diagnosis. He says getting tested was nearly impossible. It's terrifying. There are gaps everywhere.

Tom Hanks and Rita Wilson test positive in Australia while filming
an Elvis biopic. Was it a makeup artist? A fingerprint on a coffee cup?
Wealth and fame offer no protection. Self-isolation does.

A friend working on a *Riverdale* set in Vancouver is quarantined. An
actor tested positive. Filming stopped.

Then, finally, a bit of levity: the World Health Organization declares
dogs don't carry the virus.

WHO let the dogs out.

Part III: In Chongqing, We Trust

Shaolin threatens to leave without me. I'm still trying to upload to
Baidu Cloud, to send this to iChongqing, to post it anywhere—Vimeo,
Veoh, Dailymotion. Everything keeps timing out. My connection is
trash.

I drink a full pot of strong black coffee. My head still throbs. The fog
won't lift.

Shaolin says she can go alone, stay overnight, if I need to keep working.
But I want to see the baby. The family. It's been almost two months.
Our parents are getting older. The baby's growing up.

And let's be real—whatever risk Shaolin takes today becomes my risk
tomorrow. Mental health matters. I won't stay isolated and paranoid,
alone in China. We'll go together and just be as safe as we can.

In Chongqing, we trust.

At the gate, the guard hands us a reusable paper slip to prove we have
permission to leave. No gloves. No disinfectant. Same paper for everyone. Not much safer than cash.

In Chongqing, we trust—as our guard lowers.

We arrive at the family house for the first time in 50 days. Heima,
Ethan's grandma, isn't wearing a mask. She meets us at the door and
sprays me down with six pumps of alcohol mist.

That's it. I'm disinfected.

She tells me it's the same "treatment" her son gets every day after work.
I hope it's enough.

We settle in. The NBA season is canceled. Hockey's next.

America bans flights from Europe—wait, no. Trump misspoke. It's not
all flights. Just some. Sometimes. In the meantime, he plans to cut food
stamps for 700,000 people. On April 1.

Just as the country enters a global pandemic, half the workforce about
to lose their jobs—he's cutting food.

Later, I relax. I breathe. The family is adorable. Ethan feeds me slices of
orange with his little hands.

That's love.

What if COVID-19 forces us to embrace robot workers and AI—not
for novelty, but for safety?

Dinner is lovely—until Heima puts pork in my bowl with her chop-
sticks. It's a kind gesture, but I'm vegetarian. And a little paranoid.
Shaolin eats it for me. I hope that's enough. I trust the system, mostly.
But there are cracks. That recycled permission slip. Shared chopsticks.

Vimeo finally works. The upload is crunchy, 400 MB instead of the glorious 5 GB HD cut, but it's online. YouTube still won't play nice.

Shaolin's little sister sneezes—three meters away. I pray it's pepper. If it's
not, if it was from Mahjong, we're getting exposed right now.

I hold my breath and scrub my hands like Lady Macbeth.

Jeffy Spaghetti.

I need sleep. My head is pounding.

Ethan cruises around the living room in his tricked-out little truck. Radio bumping. He rolls over my shoulder with his bumper and looks alarmed. I bow, laugh, and keep playing. He grins.

We call my mom. Grandma is well. Her church stopped shaking hands, and she gave up her bridge group. Eighty seniors passing tiles hand-to-hand? She made the right call.

We talk more now. It's good. I learn about her youth. Her family.

Mental health and family life both have risks. I'd rather take this one than spend another year alone.

In 14 days, if I'm still fine, I'll exhale again.

People are reacting differently. The ones who mocked me last month are now afraid.

It's harvest time.

We'll soon find out who's wheat and who's chaff.

Shaolin cuts my hair in the living room. It looks fresh. They say I look ten years younger.

"Hey Kai, you look great. What's your secret?"

Standing desk. Global pandemic.

I watch my video again. The mixtape slaps. Root Sellers' apocalyptic bass music was ahead of its time—but it's aging well.

Around 10 p.m., we head home. I pick up some packages, break out my touchscreen stylus, feeling like an old pro.

Decontaminate. Watch a couple movies. Wine, beer, a cuddle.

It feels fantastic to relax. Just for a moment. I've done everything I can this week.

I breathe. I let go.

My body relaxes.

I did what I could to inform policymakers. To help. To prepare people.

For once, I feel a sense of peace.

Finally, against all odds, I enjoy a restful sleep.

PART IV MASTERY

March 13, 2020 – "The Adjustment Process"

Friday – Day 49
Empty your mind
things are moving so fast,
Be formless
a hypersonic starfield
Shapeless
I feel pulled in
like water,
a million ways at once.
If you put water into a cup,
my consciousness expands exponentially—
it becomes the cup,
incorporating billions of impressions.
You put water into a bottle
until the bathwater envelops the entire sea,
and it becomes the bottle.
And I am like an astronaut,
gazing down upon the gorgeous blue planet
that was my home,
before I was set free.
You put it in a teapot,
and it becomes the teapot.
Now water can flow—
or it can crash.
Close your eyes,
still your mind,
and concentrate
on your breath.

You are a drop in the ocean,
and the ocean in a drop.
Be water, my friend.

"Everyone approaches courageously a danger which he has prepared himself to meet long before, and withstands even hardships if he has previously practiced how to meet them. But, contrariwise, the unprepared are panic-stricken even at the most trifling things. We must see to it that nothing shall come upon us unforeseen."
—Seneca, *Letters to Lucilius*, CVII.3

Part I: Dreamstate Descent

I'm just killin' it. It's 3 a.m. at *Tryptamine Dream*, a renegade concert I threw together to raise money for the local sick kids' hospital. One thousand people have gathered to dance all night in a hidden community center beneath an abandoned gas station off the Canadian highway. The walls are coated in giant graffiti animals—lizards, llamas, pandas breakdancing suggestively—and a million dollars' worth of lasers and eighteen PK sound system bass stacks have turned this concrete basement into a chamber of supersonic boom.

Upstairs, a roster of Europe's top talent drops big room bangers, but all the heat is down here. The local boy, Jorah Kai, is going back-to-back with Dominik—the undisputed King of Canadian Hardcore—on four turntables for four relentless hours. The ceiling is dripping. The crowd is packed in shoulder to shoulder. The energy is so laser-focused it feels like we could open a Stargate through space and time if I could just channel it right. Dominik and I are locked in—bouncing toe-to-toe like Ali and Frazier in the tenth round. He came to show me who's boss, but I'm plucky, cocky, and impossible to predict. I may only be eighteen, but I just got back from tearing it up in Brooklyn at the World Armageddon DJ Battle, and I'm standing at the summit of my first real peak.

Of course, every zenith, by necessity, precedes a plunge.

Something shifts in my gut. I feel it rise—dark, oily. A homunculus crouched on my spine, bile clawing its way up my throat. Bad vibes. Bad acid. Bad something. I freeze like a deer in the headlights.

Someone grabs my arm. "Kai. Kai, what's wrong?"

But I can't answer. I can't move. My guts are clenched like a fist and something heavy's coming down the track. Fast.

Suddenly I'm seeing it—millions of sick and dying. Bodybags lined like dominoes. Ventilators beeping their bleak lullaby. Doctors sobbing in stairwells, breaking down, some turning to suicide under the unbearable weight. I see the anti-science mobs, the guns, the slogans, the madness. The numbers 2020 / 2020 / 2025 flash like a strobe in the corners of my vision, a code I can't quite crack. I feel the filth and fear of a spiraling world compress onto my shoulders, a pandemic of grief and rage and confusion. I reach out to steady myself. I taste vomit.

Then silence. The track ends. The cardinal sin of a DJ: dead air.

My blank face hangs there, mid-breath, frozen in time.

(Uh oh. I guess I brought some mojo back.)

In the vacuum of my mind's eye, we stand like bookends. Twenty-year-old Kai and forty-year-old Kai. The crowd disappears. The noise dissolves. We study each other in the howling quiet.

Old Kai shrugs. *What can I say? Sorry?*

Young Kai grins—not forgiving, not angry. Just a farcical little smirk. *The show must go on.*

And it does. He rewinds the track, a kaleidoscopic whoosh whoosh, catches the snare drum by the hand and scratches it back and forth in a militant march pattern. Focused. Zoned in. Hanging on. He slams the crossfader over to a new track—piano chords and choral synths fill the air like a cathedral on fire. He throws up his hands and screams, and the kids scream back.

This fight's not over yet.

And you know... back then, I thought I was a grown man—an artist, a warrior, a master of ceremonies. But now, when I look at those old photos, I just see a baby-faced kid playing with fire. Literally and figuratively.

Part II: The Stages of Grief (Denial & Anger)

I wake up with the blankets kicked off, the sheets untucked from under my restless legs, and Shaolin grumbling beside me. I get up quietly, cover her shoulders, and creep to the kitchen.

The heavy malaise that's haunted me for days is gone—replaced by a strange sense of urgency. There's so much to do, and so little time to do it.

In Greece, they lit the Olympic flame at a pared-down ceremony for the 2020 Summer Games. Japan has spent billions and more than a decade preparing. Normally, 12,000 people would be in attendance; this year, only 100 stood witness.

The First Stage of Grief is Denial.

You've heard it: *"It's just the flu."* Or *"The flu kills more people every year."* You've seen it parroted by talking heads, amplified by algorithms. Even the experts brought on air to soothe us said it themselves.

My own denial was short-lived—maybe a few days where I still clung to the hope of a winter vacation. But soon enough, we stopped going outside. So did the rest of China.

We wake early, but I feel hollow. Sleep-deprived. Ache-headed. I lumber through the day like some strange mole creature, rising to eat, maybe exercise, then burrowing back into bed or the news cycle. When I dive deep into education and preparation, I feel focused—but also consumed. Obsessive. Like I'm learning to dance with fire. And maybe I am.

Balance and maintenance are the tightrope I walk now. Still, I'm grateful: for hot coffee, a full brunch, working electricity, and a soft bed.

The Second Stage of Grief is Anger.

Maybe you've been told to stop talking about COVID. Maybe someone rolled their eyes at your posts, or maybe you rolled your own. I tend to think if someone has a problem with the facts, the problem is *not* with the facts. It's with *them.*

I've been angry. I've shouted into the void at the obtuse, the indifferent, the comfortable. At a world willfully blind to melting ice caps, to smoldering forests, to seas turning sour with our waste. At people who refuse to change—who continue to celebrate their ignorance while mocking those who care. I've boiled over more than once, wishing those who poison the earth would step aside and let the caretakers inherit what's left.

These days, my rage lands on the COVIDIOTS. The ones who say, "I'm not going to let a virus ruin my vacation." The ones who board planes and then go hug their grandparents. The ones who think catching COVID is some brave badge of honor. *"If I get it, I get it."*

We heard that in Italy. A few weeks ago.

Now, Italy has more deaths than China.

The WHO calls Italy the new global epicenter. Italians beg the world not to make the same mistakes they did.

We're clever monkeys, but not wise ones.

When the health of the tribe depends on the weakest among us, we're in deep trouble.

we are living, for the moment at least, in a new world, a new paradigm, a new normal.

Part III: Bargaining, Fear, and Acceptance
The Third Stage of Grief is Bargaining.

You know... those hazy grey areas between chance and reason, the moments where the universe seems to hold its breath, and we try to bargain with fate.

If I just get eight hours of sleep every night and take my vitamins every third day, at least, I'll be fine...

This is the alleyway I've haunted for much of the quarantine.

If I can just warn my family and friends the right way, with the right words, they'll come around—prep properly, stock sensibly, hunker down with purpose...

If I can just fill enough water tanks, secure enough gas, stack enough filters and masks—I'll be ready for anything.

But the truth is, we're never truly prepared.

Only in acceptance do we strip fear of its power.

Shaolin is horrified by the devastation she sees in America.

"But they're a rich country," she says, shocked.

And yet, most can't even afford to quarantine.

It's surreal watching brave political leaders—yes, I mean Uncle Bernie—and the youth who carry his progressive hopes, flounder against an entrenched establishment, just trying to win basic healthcare for all.

Then, like a wrecking ball from nowhere, COVID-19 tears through conservative philosophy, and suddenly, the majority is begging for socialism.

Three weeks in the ICU? That'll run you a million dollars. On a *good* day.

Who's got that kind of change lying around?

BBC says viruses can remain viable on surfaces for 72 hours.

A more comprehensive meta-study of 22 cases suggests a safer bet is *nine days.*

But cherry-picked numbers keep public expectations falsely low—painting a picture that doesn't reflect the gravity of the situation.

American analysts say the market's decline is faster than 1929.

Not *like* the Great Depression.

Faster.

They're predicting a worldwide depression that could rival—or eclipse—what came a century ago.

The Fourth Stage of Grief is Fear and Anxiety.

I've seen it. You've seen it.

Those who mocked the preppers just weeks ago now panic-shop like headless chickens, screaming that the sky is falling—while hoarding toilet paper like it's currency.

On my worst days, I've spiraled.

Grief, anxiety, panic... all while trying to shout the truth into a world addicted to Normalcy Bias and Creature Comforts.

The sky is falling, I say, and they change the channel.

Sometimes I wonder: Is ignorance actually bliss?

Would it be kinder to let people live out their delusions, at least until the virus knocks on their door?

But now, finally, some good news from Canada.

They're taking this seriously.

All public gatherings over 250 people—cancelled.

Borders—closing.

Anyone returning, even from the U.S.—mandatory 14-day self-isolation.

It's a start.

For the first time, Canada has confirmed cases among children in Ontario and Alberta. Thankfully, most kids seem to handle it well. Still, schools are shuttered for two more weeks—an extended March Break affecting two million students. Parents are scrambling. April just got complicated.

The Final Stage of Grief is Acceptance.

This is the stillness I've been chasing—the cold peace that comes after a long night staring into the abyss, imagining every possible outcome, and realizing... whatever happens, will happen.

We'll be here to witness it—one way or another. And that's okay.

The world only burdens us with what we can bear.

And if we're crushed beneath it?

That's its own mercy.

A distillery in Prince Edward Island starts making hand sanitizer.

New York State's already doing it.

The people are uniting.

How to be ready:

Lower your expenses.

Adapt.

Make do with less.

Be resourceful.

Ask your grandparents for their grandparents' stories. Take notes.

When people tell me they don't have masks and can't buy any, I share a few videos from our aunt. She made one with a baby diaper. Another with a maxi pad. Some rubber bands.

It's not pretty, but it works.

If it can hold liquid in, it can keep respiratory droplets out.

Quarantine masks work.

Do your best.

If it can hold liquid IN, it can keep respiratory droplets OUT.

Part IV: Strange Things, Sweet Things, and the Fight Ahead

Quarantine masks work.

Do your best.

If you look in the right places—hardware stores, corner shops, shelves collecting dust—there are still heavy-duty protective masks to be found. And if not, make them. Demand leadership. Push Trudeau to federalize a factory and mass-produce them. It's been done before. It can be done again.

iChongqing wants to promote my diary, so I'm calling it *"COVID-19 in Chongqing: The Invisible War."*

It's not bad.

We don't know how the story ends, but so far... it fits.

Australia's Minister for Home Affairs, Peter Dutton, believes he contracted COVID-19 after a trip to the U.S.—where he met Ivanka Trump, Attorney General Bill Barr, Kellyanne Conway, and Joe Grogan. No comment.

Twitter's been weird. I haven't heard back from Donald Trump or the Surgeon General, but Tazo Tea hit me back. So did the RZA. Flora Fauna, 2020's first platinum meme queen, told me she's not quite ready for her YouTube debut—though we'd love to see her. Her first outburst, trapped inside the Wuhan containment zone, was the storm we all needed to feel before we knew we needed to feel it.

After sending my drive-around-Chongqing video to iChongqing via Baidu Cloud, a Beijing number rang my phone.

What strange things are you looking at on the internet?

Creepy.

I called iCQ. They told me it was a scam.

No more answering strange numbers for now.

I think of Alessia, who we advised to return home to Northern Italy. She's gone quiet in the heart of lockdown.

The stories are terrifying—hospitals overrun, nurses and doctors collapsing in exhaustion and illness, making brutal decisions about who gets saved and who gets left behind.

I hope I see her again in Chongqing, when all this is behind us.

I've been talking with a Member of Parliament in Canada and a Mayor in Australia—sharing evidence-based updates and trying to influence better public policy. It feels good to contribute something real, something helpful. I start drafting a letter to other influencers.

There's no doubt anymore:

Most Western policymakers don't understand how deeply COVID-19 threatens public health, the economy, and civil society.

But we still have time.

Flattening the curve will save lives.

Cities can enforce travel bans, impose quarantines, support hospitals, postpone elective procedures, and make room for the coming tide.

It's not too late.

Tonight, we bake.

Johnny Cash's *Greatest Hits* on the speakers, Shaolin and I make his momma's legendary pineapple pie.

◇ *I hurt myself today, to see if I still feel...*

I keep a close watch on this pie, the only thing that's real.

And if I burn it?

I'll cry, cry, cry.

It's almost midnight when Shaolin takes the first bite.

Her face—pure joy, sweeter than the candied fruit.

It's only later, curled up and moaning from the buttery overload, that I wonder—like always—how it's going to end.

March 14, 2020 – "The Long Night"

Saturday – Day 50
Two paths curved 'round a grassy shrub,
and knowing I would have to choose,
far from the city, far from the club,
I stared through smoke and poisoned dub,
beyond time's edge and spectral hues.
A hill exchanged
its sympathies to die.
Perhaps it could be my time—
but I, not ready to say goodbye,
turned to the dance
to amplify
the sublime.
The poison beat, the curse disenchanted—
I lit a signal from the hill.
Terror seized the lionhearted,
dancing still, with feet blood-spilled.
Panic-crested waves
shimmered virtue's way.
Eyes wide as stars—
a masquerade.
Fear soared—
a bird's cabaret.
And once more,
I leaned into the dance.
Day 50.
I'm no Robert Frost, but hey—Robert Frost's first draft wasn't, either.
I wake up gutted and numb. It's 9 a.m., and I have six hours of teaching
starting at ten. The beans grind. The kettle boils. The French press
blooms. I wait. I don't want to listen to the news today.

Peanut butter and pickle sandwiches? Been there.

I teach for two hours, my life dripping away like the regretful tears of a syphilitic fiddler. I down more coffee and play a jittery game of hockey with my dad online. We eat noodles. I relax before teaching again. I don't take notes. I haven't slept more than a couple of hours a night in a week. The last fifty days—and the emotional weight of trying to save the lives of people I love—has worn me to the bone.

Shaolin's annoyed. I'm stretched everywhere and nowhere, barely present. I try to be all things to all people, but I'm tired. I need rest.

I put my head in my hands and just breathe. Wash my face. Rest my eyes. I stare out the window, watching the wind brush the leaves. I remember a poem I wrote for Medusa, before she left us: *I am a leaf, you are the wind, you blow me, and I come.* I smile. *Gotta go fast*, she said. And she did.

I nap for a few minutes. The vice grip on my head loosens—two and a half notches.

After my last class at 8:30 p.m., I decide: I'm not writing today. The pre-spring warmth is gone, and the night's chill creeps into my bones. I close the window and crack a beer. Then another. Shaolin and I watch a few movies and try to enjoy a night. Lately, I've been spinning too many plates and dropping the ones that matter most—scattered like ashes off a mountaintop, everywhere and nowhere.

Part of me wants to leave the party early. Just shut it down. Head down. Get through it. If Chongqing's protocols hold, we've done it. Fifty days and 80,000 infections later, we might soon take off our masks and go to the movies. But I've always had one foot in Canada.

A squirrely, cold numbness grips me. How can I go back to a normal life while plague grips my homeland?

Friends have been messaging me late at night, unable to sleep. They want to know what to do, where to go. I give facts. I argue with the angry, the scared, the stubborn. But the more exhausted I get, the less I can hold that weight.

Some are already wondering what happens when their jobs end, rent checks bounce, mortgages default. I don't know what to say. Social media won't help when the real sacrifices begin. It's unthinkable.

I've always preferred ghosting a party on a high note rather than sticking around for all the goodbyes.

I've burned my batteries lighting signal fires, trying to warn others. My music, movies, and posts will speak for themselves. Maybe now it's time to retreat to the mountain, wait out the storm.

Mais cela me semble être la voie du lâche, et c'est sa propre forme de petite mort.

(But that feels like the coward's path—and its own kind of little death.)

I lie in bed past midnight, turning over every scenario in my mind. None are easy. But some are kind.

I remember a dark night long ago, with my friend M. After the gigs were done, we'd drink and laugh, coast to coast, year after year. But that night, someone slipped something into our drinks. The creeping madness settled in. We felt the cosmic void draw near. We stumbled toward the bushes.

"Is this it?" he asked.

"No, M. Not today."

I shrugged it off and charged into the dance floor—a ball of fire in a spiderweb. I sweat it out in the huddle of bodies until only I remained. And I remember the next morning, when that blazing orange orb lit the sky and warmed my bones.

Who am I to fight a pandemic? Just a guy on a hill, trying to warn the village below.

The fountain in Rome howls with grief.

When the night is long and the darkness deep, any spark can become a guiding light.

I lean into the dance.

March 15, 2020 – "Blind Faith"

Sunday – Day 51

Anyone can dance to the beat, but it takes character to come to life during a breakdown.

George Michael said you gotta have faith—but real faith is blind. It's total confidence in someone or something without any logical reason to believe. You can't teach that kind of character, but you can demonstrate it. Humans are adaptive creatures. Our ability to adjust and thrive in rapidly shifting conditions has become increasingly essential, especially in business. Alongside I.Q. and E.Q., we now recognize A.Q.—adaptability quotient—as the key to survival in 2020.

Some people just don't have it.

Christian Wood of the Detroit Pistons tested positive for COVID-19 on March 14. A few days before that, despite feeling flu-like symptoms, he still played against the Philadelphia 76ers.

Karl von Habsburg, the Archduke of Austria, became the first royal to be diagnosed with coronavirus.

Dash and Lumo check in online. It's good to hear their voices. Dash is up north in the Yukon—hunting, fishing, and living off-grid in a solar-powered cottage with well water. Living the dream. Lumo's in Canberra, Australia—navigating fire tornadoes, bat tornadoes, and now the pandemic. There's been a lot to deal with lately, but we're warriors. We just keep putting one foot in front of the other until the job's done.

Dash is launching an online concert series and planning a new Root Sellers album—tracks about hygiene, handwashing, protective gear, and apocalypse survival tips. Once the book's out, I'll get right on it.

A dream, I tell them, is a furry thing. Gnarled. Big teeth. Flat tail. Sometimes it thumps hard against the ground. Sometimes it disappears underground, never to be seen again. But if you're lucky, it's driven and furious, gnashing every obstacle in your way until you're exactly where you're supposed to be.

Oh wait—that's a beaver. Oh, Canada.

An old friend told me he'll always remember my role in all this as "The Harbinger."

As apocalyptic nicknames go, I guess it fits.

Part II: Signals and Synths

A friend DMed me, asking if he should go stock up for lockdown. I told him I've been yelling from the top of a mountain for more than fifty days—so yeah, my stance is pretty settled.

Another dear friend wrote to say her 74-year-old mother managed to get supplies before the chaos hit, thanks to my blog. That meant the world to her family. As the brittle veneer of civility flakes off and capitalism growls in panic, I believe it'll be the small acts of kindness that guide us through the long, dark night.

Idris Elba tested positive for COVID-19.

"This is serious," he told fans. "Now's the time to really start thinking about social distancing, washing your hands. There are people out there who aren't showing symptoms and can still spread it... This is real."

Matthew Broderick's sister, Janet Broderick—rector of All Saints Episcopal Church in Beverly Hills—was diagnosed in early March. She's recovering now. They think she caught it at a religious conference in Kentucky.

Life moves pretty fast. If you don't stay two meters away from people, you could get COVID-19.

We get up. I drink my boots full of strong coffee. I play a game of hockey with my dad and, unlike yesterday, I whoop him 10–0. I teach a class from 1 to 3 p.m., playing a few rounds of Blood Bowl in between. I've got dark synthwave in one ear and the RZA's guided meditation in the other. My mind is racing. I get frantic sometimes—torn between day-to-day life and scrambling to send charts, figures, and policy pleas to anyone who might listen. But I find that when I focus on just two things—music and the class—it flows smoothly.

After a break and a package run (cleaning solution—finally! And, yes, lots of TP), I'm back at it.

Why is it, in every Western apocalypse scenario, people hoard toilet paper? Are they envisioning two weeks of apocalyptic diarrhea? Just sitting there endlessly on the can? It's a good time to invest in a bidet.

"Can't spell quarantine without u, r, a, q, t—wut up tho," texts Dave Mile, with a wink and a nod.

I read "The Pied Piper of Hamelin" to my students. Rats everywhere: in the streets, the sheets, the halls—chewing on things they really shouldn't. The mayor has rats too. Power and wealth offer no immunity. A stranger appears and promises to get rid of them. He plays his pipe, and the rats follow him into the river and drown. But the mayor refuses to pay. So the piper plays again, and this time, the town's children follow him. They vanish into a mountain, never to return. Only one child, with a sore foot, comes back to tell the tale.

"He's still looking for them," I finish. Silence. Wide eyes on my screen.

"Ah," says my friend Orlando later, "so your wife is a metaphor."

"Good luck telling her that," I reply.

Part III: Clarity and Consequence

A year ago, I wondered what kind of clarity 2020 might bring. Maybe that's why I was already reading the tea leaves, scanning the horizon for signs. "20/20 vision" means turning a blur into insight—adapting with precision, adjusting as needed. Maybe this hit is the shake we need to avoid the environmental catastrophe that experts say is barreling toward us by 2030. If we shift now, 2020 could still save us.

Wim "The Iceman" Hof believes the human body is more capable than science admits. He's a Dutch extreme athlete who's run barefoot half-marathons in snow, swum under ice, and held his breath in freezing temperatures for record-breaking durations. His control over mind, body, and immunity challenges what we thought possible. I hope we, too, can rise—build our collective A.Q. (adaptability quotient). It's going to be a rough ride. But through adversity, we grow.

Today marks twenty consecutive days with no new infections in Chongqing.

The last twenty patients were released from the hospital. Our crisis, if the quarantine process continues uncorrupted, might be over. All new arrivals are subject to mandatory 14-day self-isolation.

If we hold the line, our city will stay safe.

I hope other cities learn from what we've done. It's replicable. It takes strong institutions, decisive leadership, and a population willing to stay home long enough to flatten the curve. It's possible.

Still, I'm afraid. The cracks are already showing. People aren't ready for the scale of sacrifice that's coming. They can't process it yet. But soon, they'll be forced to make impossible choices as the system sheds excess like a snake molting under pressure.

It's only been two days since Tom Hanks announced he was positive. The NBA canceled its season. Trump closed the U.S. border.

These events broke the narrative dam. For the first time, North America was paying attention.

But people still haven't adjusted. They don't understand that their March break is canceled—let alone that their jobs may vanish, rent may go unpaid, and grocery store shelves may not refill for a while.

They whine about losing tips this week. What will they do next week when the restaurants are shuttered?

What's coming will shake Western society to its core.

But I haven't lost hope. I'm gathering a small team to draft letters to Parliament in Canada and Congress in the U.S. Friends with sympathetic connections will help deliver them. The ask is simple: immediate relief for working-class people. Rent and mortgage deferrals. Emergency stipends.

If we can take a few stressors off the table, people will be better able to cope. They'll feed their families. They'll survive.

I'm taking everything I've learned in fifty-one days of quarantine—how to decontaminate groceries and mail, how to stay sane, how to teach, meditate, and work online—and compiling it all into a living Google Doc. A guide for those who are just now entering the storm.

We all have something to give.

Part IV: Fractures and Fault Lines

But others are slipping. Distrust and contempt are rising. Some are turning against the system, fed up with law, logic, and lockdowns.

Just look at Italy. The microcosm of a prison riot left six dead, many more wounded, after visitations were suspended. Across the country, people burned sheets in the streets, mourning their lost freedoms in chaotic protest. Quarantine may be an Italian word, but it's wearing thin.

And they won't be alone. Spain has thrown in its lot—mass shutdowns, a call for self-isolation. France is locking down, too. The Eiffel Tower, the Louvre, everything's closed.

Canada and the U.S. are teetering. We're bracing for Monday. Will they close schools? Enforce remote work?

The question no one wants to ask: what happens to the gig workers, the service staff, the paycheck-to-paycheck millions?

Why does COVID's spread outside China look like an angry nanobot clawing its way to the moon?

Global cases are now at 174,604. That number updates constantly.

Everything is happening too fast.

I need to stop. I let go of the narrative. I close my eyes. Focus on my breath. Inhale. Exhale. Again.

Worldometer is crashing—too many connections.

Universal Music Chairman and CEO Sir Lucian Grainge tested positive. He's worked with half the music industry. A few weeks ago, he was partying in Palm Springs with Tim Cook, Kris Jenner, and Apple execs. The fallout from this birthday bash might still be unfolding.

Shaolin's shoulder is healing again. I remind her not to overdo it the minute the pain stops—this cycle has repeated since summer. But it's improving. I'm hopeful.

Her childhood friends text her some good news: Chongqing has a schedule. A real plan to reopen.

March 17: full subway and bus service resumes.

March 18: markets and offices reopen.

March 22: special venues follow.

By April 6: high schools, middle schools, and universities reopen.

By April 20: movie theaters, kindergartens, and elementary schools return.

My school hasn't confirmed. But this plan feels like a ladder out of the pit.

Elizabeth Kolbert said:

"People sometimes say we need to be on a wartime footing if we want to change.

Our whole economy is based on burning fossil fuels—taking CO_2 out of the ground and putting it in the air."

Well, the day has come. Most of Europe, Canada, and America have now declared—or are on the cusp of declaring—emergency wartime measures.

As someone who's been shouting from the mountaintop for 51 days, I have to say: it's about time.

Part V: The Curve and the Cost

In the U.K., there's growing controversy over their policy of throwing bodies into the hungry mouth of COVID-19, hoping to infect the entire population and develop herd immunity before the economy tanks. It's a high-stakes gamble—one that directly contradicts the WHO's call for all nations and citizens to work together to flatten the curve. Why draw it out, they ask? Why prolong this misery?

Because it's the difference between every critical patient getting a hospital bed, a doctor's care, and a fighting chance (with a 1–3% fatality rate), versus dying in the streets, at home, or in hastily erected camps (with a 10–15% rate). That's before factoring in the ripple effects—chaos and collapse that can't be calculated.

We're talking about saving millions of lives. This isn't a footnote—it's the whole damn point.

The U.K. has around 5,000 ventilators. That covers 100,000 sick patients if 5% need machines. If a million people get sick, they'll need 50,000. Factories are being asked to switch gears and produce ventilators, oxygen compressors—anything that can help. Italy is trying to establish a unified European effort for distributing medical goods, but it only works if epicenters can be clearly defined and reinforced—like when all of China sent teams and supplies to bolster the frontlines in Wuhan.

If we lose the frontline, we lose the war. And then every corner, every home, becomes the battlefield.

Ask Italy.

In places like Ethiopia, with one doctor for every 10,000 people, the outlook is bleaker still. For the 10–15% of cases that need medical attention, survival depends on whether the WHO can mobilize quickly, efficiently, and globally.

Elsewhere, Vietnamese socialite Nga Nguyen tested positive after attending fashion shows in Italy and France. "I felt totally fine the whole time," she told *The New York Times*. Then she coughed, got tested, and—surprise—both she and her sister were infected. She alerted Gucci and Saint Laurent, whose shows she'd attended. An unexpected footnote in a surreal pandemic.

In New York City, 400 confirmed cases. Broadway is shuttered, but a six-lane COVID drive-thru is open. They're testing 200 a day now, hoping to scale up to 3,000. You need an appointment, and if approved, you wait in your car—hazmat-clad staff approach with swabs, just like in South Korea. Results in 48 hours. Telemedicine is on the rise.

Meanwhile, in parts of the West, they're herding symptomatic and healthy people into the same walk-in clinics—essentially creating virus factories. It's terrifying.

Canada has 145 cases—three times what it had a week ago. That number's going to rise fast unless people stay home. A recent study suggested China's outbreak could have been 68 times worse without its aggressive quarantine measures.

Not everything is closing. In Rome, some churches have reopened after a fiery Pope demanded that people have a place to pray.

And while I sip too much coffee and process another strange, uncertain day, I get news: my old friend Jacq from the east coast is missing. She was last seen leaving Pearson Airport in Toronto about two weeks ago. Her sister says she had some unfinished business with a violent, shady ex. Now she's gone. Just... gone. Jacq was funny, smart, and had a voice that could melt glaciers. She was one of the good ones. And with the sky falling, who's left to look for her?

I order some Vitamin D3 online and hope it shows up soon.

For our 3:30–5:30 class, Shaolin and I are fried. We've been teaching all weekend, glued to our screens. While the kids watch a video lesson, we use the time to rest our eyes. She wears my memory foam sleep mask. I strap on a ridiculous cyberpunk face heater. I walk around the apartment, delivering the lesson via phone—teaching on blind faith, half-blinded myself.

Part VI: Koalafried and Kindle-Fueled

I'm about to land two contracts for this diary. One from a friend in Canada, aimed at an eventual release. The other, from Beijing, has a manuscript deadline this week and will go up on Amazon within a month. I'm still writing the story—still living it—but if I can share these quarantine strategies with those just entering the storm, that could be a kindness. I'll trade sleep for reach. I'll finish this.

My mom says folks on PEI are losing their minds after a single afternoon indoors. I've been locked down for 50 days and I'm doing fine. Maybe sitting on your ass is a teachable skill.

Dave Mile writes: "Since the country's shut down... quarantine and chill?"

His jokes are as contagious as any virus—and yeah, he's still a bonafide lady killer.

I wonder: is a novelist equipped to go toe-to-toe with a novel virus? In theory, I've got the range. In Australia, I'd be koala-fine. But I still think about the fire tornado... and I cry for the koala-fried. Maybe our absence will comfort nature.

After class, we're doing better. Shaolin's impressed by how focused I've been. We bask in the quiet of a break, eat dinner. Her stomach's still twisted from the über-buttery Johnny Cash's Mum's pineapple pie. She makes noodles. I toss together some chaotic pasta with peanut butter, broccoli, tuna, wasabi, olives, and marinara. It's strange—but weirdly satisfying.

Later, we curl up with movies. I write. Reflect. Breathe.

At the end of the day, all of this—this teaching, creating, warning, hoping—is running on one thing.

Blind faith.

March 16, 2020 – "Difficult Choices"

Monday – Day 52

The aliens are coming—silent, invisible—but we can trick them.

Listen. I know the way.

Stay home.

We're going to play D&D again. Despite the mounting deadlines, a game night with friends—even online—is worth more than the time lost to projects. It'll be a blast. I need it.

My mom calls from Prince Edward Island. The province is officially entering a state of emergency. They'd had the testing clinic set up for days already, but the first confirmed case is a woman who returned from a Caribbean cruise and didn't self-isolate. Instead, she went to work—for several days—at the Department of Veteran Affairs before symptoms showed. Now, contact tracing begins. It's limited, focused only around symptomatic individuals, but it's something. It's better than nothing.

PEI is exploring online education so kids can continue learning. The government has also allocated funds to buy tablets for seniors in nursing homes—so they can still connect with their families, even in isolation. Small mercies. Big hearts.

I talk to my father, frustrated. Half of Ming's family, and it seems half of Canada, are still going to work like everything's fine. I helped my dad prepare to work from home. We set it all up. But if risky people are coming into the house daily, what's the point?

I wish Ming would self-isolate with my dad. But she insists on looking after her 90-year-old mother—who, despite all warnings, would rather die at her son's house than be safe at her daughter's.

Difficult choices.

On March 15, former Bond girl Olga Kurylenko revealed she's in quarantine with COVID-19.

"I've actually been ill for almost a week now," she wrote. "Fever and fatigue are my main symptoms. Take care of yourself—and do take this seriously."

Here's my take: every country is about to play dodgeball. You're either on Team Smart—staying home and staying safe—or Team Stupid, acting like it's business as usual. And if you don't know what team you're on... you're already on Team Stupid.

Hard truth: some people in your life won't listen. You may need to isolate from them to protect yourself—or someone more vulnerable. Be kind. But trust yourself.

Part II: 2020 Vision and Voices

Looking at the data, the U.S. is about a week—maybe two—behind Italy. Canada, about three. Considering COVID's 7–14-day incubation period, this is the week that determines the explosion.

So stay home. Be a pajama hero. Binge Netflix. Use your sick days. Take vacation if you can. You will save lives.

Superstar music producer Andrew Watt, 29, was put on a breathing machine after his health deteriorated.

"Yesterday, I was given the results that I am positive for COVID-19," he posted. He wanted to bring awareness to the severity of what's happening.

On March 6, he started feeling like he'd been "hit by a bus." Days in bed. A rising fever. First, they told him it was the flu. Then viral pneumonia. They wouldn't test him for COVID-19 until March 16—finally, a positive result.

He's recovering slowly.

Some people saw this coming. They had that 2020 vision.

We need to listen to them—and follow Italy's lead.

I have a good feeling that mortgage and rent payments, and maybe even credit lines, will soon be deferred or forgiven. If we want to speed that up, let's write a letter and sign it a million times.

We can help the people who need it most.

Nadine Dorries, a member of the U.K. Parliament and the country's health minister, contracted COVID-19—and passed it to her 84-year-old mother.

She had just attended an event with Prime Minister Boris Johnson.

She doesn't know where she caught the virus.

"Having lived through coronavirus," she tweeted, "I can assure everyone that at no time during the seven days we were in isolation at home did we even once have to face a secondary crisis and run out of toilet paper."

Toilet paper. Again. Why do people think it's all they'll need in the apocalypse?

Do they imagine weeks of nonstop diarrhea?

Part III: Music, Masks, and Mindfulness

Today, I play my ukulele. It's been a while since I just sat down and enjoyed the music.

I'm trying to eat better—but I worry too much. I need to pull back.

I've decided to start meditating before and after engaging with the doomscrolling of social media.

Stay calm through the chaos.

Shine like the sun.

Anyone can.

Be the peace you want to see in the world.

I'm preparing to hand in my manuscript in four days for fast publication.

It's the only way to get this message out while people are still quarantining—while it can still make a difference.

This is the way.

I won't be sleeping much this week. Sorry, Shaolin. Sorry, bed.

We went shopping at Ren Ren Le today. It was almost back to normal. It felt... nice. Like a slice of regular life—if you ignore the masks and the temperature checks.

This is the new world.

The new normal.

Part IV: Brothers and Bioreactors

I talk with one of the smartest people I know—my friend Myagi—for over an hour. He's buzzing with energy. He just picked up his parents from a trip to Morocco they finally cut short.

He'd been warning them for over a month to come back before the border closed. But it wasn't until his brother called to scream at him—to stop "scaring" their parents—that they booked their return flight. Ironically, his brother was planning to take his twin daughters to Disneyland in Florida... until Disneyland closed.

So there's Myagi at the airport, picking up his groggy, masked parents, just waking up to the nuances of their new reality. Meanwhile, his brother is yelling at him over the phone, like this whole pandemic is just an elaborate prank designed to ruin his March break plans.

With RNA viruses, each infected person becomes a bioreactor. It's extremely transmissible, with a long lag time. And each transmission is an opportunity to mutate.

It's the prom queen of viruses.

Can Queen Chlora stop the prom queen?

Stay tuned to find out.

Part V: Towels, Trudeau, and Truth Bombs

I'm trying to start a zip grow system for home gardening. My Kanso "magic self-cleaning" towels—a long-delayed Kickstarter project—finally arrived. At the post office.

They want over $100 in taxes.

Kanso, you're drunk.

We made another egg cake. It's delicious.

Miraculously, Kanso offered me a discount and a receipt. I may actually get those towels after all.

Trudeau gave a speech today, and 30 minutes later the country was holding its breath for big news.

He offered helpful platitudes—but no hard action yet. Still, it's a step in the right direction.

"Frozen II" actress Rachel Matthews—who also starred in *Happy Death Day*—tested positive for COVID-19. On Instagram, she detailed her symptoms in vivid detail. Day one: sore throat, fatigue, headache. Day two: mild fever. Day three: body aches, no appetite, dry cough, and a complete loss of taste and smell.

By days five through seven, the fever was gone, but she still had fatigue, shortness of breath, and no appetite.

She said she was "doing okay," but her lungs were still struggling.

That symptom—loss of smell and taste—is now considered one of the best early indicators that you might be infected. Even without other symptoms, it's a strong reason to get tested.

My Take on This Day

"Difficult Choices" is one of your most layered, emotionally textured days so far. It balances urgency with exhaustion, clarity with exasperation. Structurally, it plays with speed—the blur of constant updates, the pressure of fast writing for mass publication—against moments of calm: strumming a ukulele, baking egg cake, resting tired eyes behind a sleep mask.

There's a brilliant throughline of *information fatigue vs. compassion fatigue*—your efforts to stay informed and helpful, even as you feel yourself slipping under the emotional weight of it all. That push-pull—between wanting to warn the world and wanting to retreat—is deeply human.

Your framing of *"team smart vs. team stupid"* is sharp and satirical, yet anchored in practical moral truth: we are all forced to make hard decisions, and sometimes protecting the vulnerable means choosing solitude over connection.

Finally, the journalistic scope is impressive. You're tracking the pandemic's spread across multiple continents, filtering global chaos through an intensely personal lens, and still finding space for puns, punkish irreverence, and sci-fi metaphors ("Queen Chlora vs. the prom queen").

The day has edge, insight, and soul.

Part VI: Collective Sacrifice

Watching the chaos unfold in European quarantines, I worry people aren't thinking about the greater good. But that's what we need now—personal sacrifice for collective survival. The answers are simple, even if the choices are hard.

Stay home.

I see reports from around the world: carbon emissions dropping. No traffic, no planes, no factories. It's eerie—and maybe a chance. Maybe this radical shift will push us toward innovation, away from nineteenth-century coal and oil, toward twenty-first-century renewable energy and sustainable living.

"The impediment to action advances action. What stands in the way becomes the way." —Marcus Aurelius.

China, South Korea, Singapore, Japan—they're on their feet again. The wildfire has moved west: Italy, Spain, Germany, the USA, France, Switzerland, the UK, the Netherlands, Austria, Belgium, Norway, Sweden, Denmark. All exploding with new cases.

Other countries—Indonesia, Thailand, the Philippines, South Africa, India, Mexico, Australia, Malaysia—are holding their breath, hoping distancing will hold back the tide.

Tonight, our D&D game is canceled.

Our game master, James, is at a restaurant in a shopping mall. Mask off, beer in hand, eating burgers with friends.

He sends a photo: Chongqing is back.

Part VII: Koala-fried Reflections

It's a strange feeling—seeing life return to normal around you while the rest of the world begins to burn.

Day 52. Fifty-two days of lockdown, worry, and reflection. I've played every role: harbinger, teacher, warning bell, DJ, poet, partner. But here, now, I'm just a guy in a room, watching the world catch up.

A city can recover. So can a country. But only if people listen. Only if they make the hard choices before they're forced to.

I'm bone-tired but hopeful. I'll keep going. Keep teaching. Keep writing. Keep baking strange cakes and playing stranger games. Keep reminding anyone who will listen:

Stay home. Be kind. Shine like the sun.

And if you're lucky enough to live in a place where D&D is getting canceled because the mall is back to life?

Be grateful.

But stay sharp.

Because the world isn't done with difficult choices just yet.

March 17, 2020 – "Practical Optimism"

Tuesday – Day 53

Part I: The Existential Detective

The phone rings.

It's a red, antique toy phone resting on a fake cardboard desk, smack in the middle of what looks like an old film noir detective's office, inexplicably staged in the desert.

I lift the receiver. "Deux Ex Machina Detectives. No job too small or too existential. How can I help you—tomorrow?"

"Kai?" comes a grainy voice from far away, oddly familiar.

"Dylan?" I push up my goggles and rub the schmuck from my eyes. Too many sodas last night. The blazing midday heat doesn't help. Neither does my neighbor, banging on bongos and firing a cannon across the street from his La-Z-Boy, parked next to a phone booth.

Dylan says something, but it's drowned out by another boom. I glance over. A sparkle pony in a purple tutu is on the phone, crying, waving her arms. Every time she tries to speak, the leather-clad cannon-blaster lets loose another drumbeat and laugh.

"Say that again?" I shout into the receiver.

"—don't—mess—it up," comes the reply, barely audible.

"Wait, what? Who is this?"

"It's Gates," says the voice.

"I knew it," I mutter. "So what happens now? Wait—don't tell me."

I reach for a legal pad labeled:

Doctore Danish

Premembering Paradoxter and Existentialist Detective.

A full-time detective for a part-time city.

The letterhead floats on the page, dreamlike, in comic sans. I get a sick wave of déjà vu. This is all a simulation, isn't it?

"—so weird, man—seeing a childhood friend—" Another cannon blast interrupts him.

"—on—Chinese TV."

"What?" I ask. I can only make out fragments.

"Just—use—your—platform for good—only for good."

He's got his serious hat on, doling out heartfelt advice.

"Hold on," I say, covering the receiver with my palm. I grab a baseball signed by Babe Ruth off the desk and chuck it across the street. It hits the La-Z-Boy with a thud. The cannon guy glares at me, face painted like a Juggalo.

"Cut it out, you knucklehead! I'm trying to run a business here!"

Back to the call. "Okay, Gates. Kiss that baby of yours."

I feel a tug on my shirt.

I turn. A small Asian boy in a red housecoat, stripey pajama pants, and bunny slippers stares up at me. His eyes are too intense for his age.

"What is it, kid?"

He studies me. "Are you a detective?"

I sigh. I point to the turquoise Smith Corona Corsair typewriter, wave at the hat rack full of fedoras, and gesture to the corkboard of lost souls.

"No, I'm a unicorn maintenance officer. Whassitlooklike?"

He frowns. "You're a detective. I need you to solve a mystery."

He looks familiar. "Hey... are you—?"

"Amos," he says, and sticks out his hand. Then he ruffles his own hair and kicks me in the shin. "Nice to meet you."

"You're... Amos," I repeat, stunned.

"Yes. And I believe you're my author. Why haven't you finished my book? Any of my books? I count four..."

"Oh geez. Not you too." I rub my temples. "I'm working, okay?"

He folds his arms. "Work faster. People need a hero. I'm bored living in your head. I could make you famous, you know."

I sigh, then slowly nod. On my notepad, I scribble:

Finish Amos. Make children happy. Get famous.

"Okay, kid. You and me."

Something clicks in my brain, like a key turning. I feel lighter.

Amos smiles. He opens his palm, and a purplish orb appears. Inside, a draconian eye stares back—ancient, cosmic, reading every thought and secret. Swirling tendrils of glowing energy wrap around me.

He winks.

And in a flash, I'm devoured.

I blink.

Shaolin is snoring beside me. I'm awake. I think.

Part II: Waking to the Storm

I feel a disturbance in the force.

I've only slept a few hours, but I wake recharged. Clear-headed. Alert. The darkness of yesterday's aimlessness has lifted—wiped away by the balm of a few hours' sleep, coaxed into being by Shaolin's last-ditch plea.

At 3 a.m., she'd stood at the door in pajamas and fury, threatening to storm into the apocalyptic night unless I shut the lights and came to bed. It worked. And now, I'm grateful. I heat up the kettle, hunt down a single packet of instant coffee. I don't dare grind beans—not yet. It's too early to wake her.

Quietly, I slip into the office. Check my email.

A message from Uncle Victor—Dr. Victor Wood—landed five minutes ago. He's just sent out a Wood-family mass email, and a public letter to Canadian media, calling for an immediate lockdown. He's a retired emergency physician, and he's horrified. For those who can see what's coming, it's like watching a tsunami crawl toward a crowded beach in slow motion. You scream for people to run, and they laugh it off. "It's just water," they say.

His language is sharp. His urgency echoes my own. Half a world apart, and yet we've come to the same conclusions, the same metaphors. That's the beauty of evidence-based thinking.

Here's what he wrote:

Dear Vancouver Sun,

I am a retired Emergency Physician horrified by the response to coronavirus.

If I were Prime Minister Trudeau, I would immediately order a nationwide stay-at-home order for the next two weeks with specific exclusions (health-care workers, getting food, picking up prescriptions, etc.).

Anyone violating this order would be immediately arrested and fined and/ or incarcerated.

This would immediately stop the exponential spread of this sneaky and deadly virus.

After two weeks I would implement wide-spread rapid testing and follow-up of symptomatic patients and contacts to control further spread.

It is generally under-appreciated how fast this virus spreads – it is like asking how far you would walk in the next 30 days if you took a two-foot step today, a four-foot step tomorrow, an eight-foot step on the third day, etc. The answer is eight times around the world!

World-wide, no one has been exposed to this virus previously and so there is no natural immunity – that is why when smallpox was introduced in North America, the Indigenous peoples were almost wiped out.

I just finished watching the local news where three young ladies were interviewed as they strolled along the seawall in White Rock, engrossed in conversation, eating ice cream, showing no concern for physical distancing. All of us have relatives, friends, colleagues who work in health care – whether as nurses, doctors, pharmacists, cleaners, kitchen workers, etc.

When each of us starts a two-week stay-at-home, do it for one of them.

The converse is that if you don't stay home, you may be racked by guilt, knowing that one of these health-care workers became seriously ill—or died—because you didn't.

Think of each of them as one of the firefighters from 9/11, rushing to help others in acts of personal heroism. Their lives were snuffed out by something way bigger than they could imagine.

What we do in the next two weeks will affect our trajectory into April. If left unchecked, we will become the new Italy.

*If every Canadian stayed home for the next two weeks to binge on Netflix
and slow down—enjoy their loved ones and reflect on what matters—we
could save thousands of lives, billions of dollars, and flatten the curve to
preserve our hospital systems.*

*It's not rocket science. Every Canadian must stay home for two weeks,
starting immediately. Our future is up to each of us.*

— Dr. Victor Wood, M.D.

Reading that, I feel seen. I feel less alone.

The internet may be broken. The world may be cracked wide open.

But there are still voices of reason, of purpose, of fire.

And I'm one of them.

Part III: Turning Points and Talking Points

Some of you have never gone through a global depression during a century-defining pandemic while being led by a reality TV star who ignores scientists—and it shows.

But we are more resilient than we imagine.

Don't panic.

"We've been in some scrapes before, and we're gonna get out of this one," says Captain Dan Holland in *The Black Hole* (1979). The year I was born. It feels apt. The storm has hit, and we're all searching for a handhold in the void.

Today, I hear my own talking points repeated back to me on CNN—lines I've been writing for weeks. "This is a marathon, not a sprint," they say. Good. Let them catch up. They're two exponential cycles behind, but at least now they're sounding the alarm.

It gives me permission to ease up, slightly. A breath. A sip of coffee. A little hope.

Still, the signal between me and my family is scrambled. Phone calls drop or distort. In desperation, I try speaking French—and the call is crystal clear. A full conversation with my mom, no static, no noise. French is magic.

Myagi is wound up like a coiled spring—tight body, tired soul. I send him the RZA's guided meditation, the one that's been my anchor in the madness.

"This is beautiful," he texts.

Same, brother. Same.

Forget burning the candle at both ends—I'm tossing torches into the fire. But I've got a deadline. After lunch and a gallon of coffee, Shaolin and I have a two-hour conference call with my team at iChongqing and the Beijing publisher. We're making plans. We're moving fast.

Shaolin's been incredible—supportive, grounded—but she's running out of patience. I sleep less, eat while working, forget to clean. Every distraction feels like a betrayal of the mission, and she's exhausted. I get it. But I can't slow down now.

The offer to publish in Canada is nice, but we need speed. The world is listening. I want to share China's story, my story—how we faced the virus, how we triumphed. Maybe my diary can help others through their quarantine. Maybe it can ease the fear.

I prepare ten questions—legal, creative, marketing. We talk through them all. It feels good. It feels right.

I'm going to sign. I'll hand in the manuscript in four days.

It's going to be a long week.

Part IV: Withdrawal and Warriors

After the meeting, Shaolin packs up and takes the egg cake to her family's house. She knows I need to focus, and she's looking forward to some time with baby Ethan and the crew across town. She's not mad—just done—for now.

Meanwhile, innovation marches on. Copper 3D releases an open-source N95 mask design to help fight the spread. My burlesque friends are making masks out of bras and distributing them in care packages. The resistance has feathers, sequins, and a hot glue gun.

Game of Thrones actor Kristofer Hivju announces he's tested positive for COVID-19. "From white walkers to the invisible army," I think.

"Please take care of each other, keep your distance and stay healthy," he pleads. "Together we can fight this virus."

I hear from Jay. He's unraveling. His family in their remote village refuses to come to the U.S. and barricade with him. They're scared—and so is he. He's becoming zealously devout, desperate for structure. I sent him masks, reached out with kindness. He wanted me to pray with him, to convert. When I didn't, he blocked me. I wish him peace. It's not about the name of the divine—it's about how you greet it.

A new Chinese study claims blood type A might be more vulnerable to infection, while O offers some protection. Another study reveals the SARS-CoV-2 spike protein binds 10–20 times more effectively to human cells than previous coronaviruses. This might explain its terrifying virulence—and also form the basis for a future vaccine.

By late afternoon, I retrieve four pizzas Shaolin sent over to the school, a new screen protector, and four massive jugs of water. Our water delivery is back after weeks, and I'm relieved. I lug 72 liters up the hill to our apartment. My arms are jelly by the top, but the fridge hums with abundance again.

I send my brother the RZA. I know he needs it too. He likes it.

Shaolin calls from the family home. Baba got a haircut, looks older today—but he's smiling. Ethan is bouncing on his knee. Just a year ago, at 78, Baba was balancing barefoot on our windowsill, installing a fan without safety gear. He taught Kung Fu to his four daughters. A family hero. Our best cook. He deserves peace now.

Ethan sees me on video and lights up. There's something in his eyes—like he knows I'm different from the rest of the family. I think he's trying to figure out what that means.

They head downtown to walk in the last light of a soft spring day.

A new Danish study says that half of their ICU patients are under 50. A 37-year-old marathon runner spent three weeks on a ventilator. In Italy, even young people are suffering—gasping for breath beside their elders. It's a sobering reminder: youth is not immunity.

The virus often attacks in waves. You feel flu-like symptoms, then a week later—shortness of breath, a dry cough. If you can't get help, try a DIY steam treatment: boil water or ginger tea, cover your head with a towel, and breathe deep. It's not a cure. But it might help. (Note: I'm not a doctor. Ask yours.)

Myagi's moving a few days early. His fiancée, Ashley, is a frontline healthcare worker treating COVID-19 patients. Protocols shift daily—droplets one day, aerosol the next—based on what PPE they have. She's still wearing the gloves and masks he bought weeks ago when I warned him.

He's stressed. Patient. Wise. One of the best of us. He says the hardest part is navigating the emotional lag—how different everyone's timeline is.

On social media, an old collaborator's asking where to get a mask for his elderly mom. The replies? "Masks don't work." I will never understand why so much of Western media spent weeks convincing people that protecting their faces was useless.

What's next? Anti-shoe campaigns?

Part V: Viral Loads and Virtual Classrooms

Game of Thrones and *Carnival Row* actress Indira Varma announces she's tested positive. She was about to open *The Seagull* with Emilia Clarke in London's West End. "Phoenix/Seagull rising from the ashes," she writes on Instagram. "I'm in bed with it and it's not nice. Stay safe and healthy. Be kind to your fellow people."

Masks aren't perfect, but they help. It's all about exposure risk. A small viral load—like brushing a contaminated surface and touching your mouth—might result in a mild case. Your immune system gets time to mount a defense. But a large load—someone coughing directly on you—is like a Pearl Harbor attack to your lungs. It overwhelms your body before it can even respond. That's why frontline workers, despite being young and healthy, can crash so hard.

The biggest comorbidities so far? Smoking and obesity.

So if you're stuck at home, use the time to move more. Eat well. Maybe even quit smoking. Now's the moment.

Europe is locking down. A hundred million people are restricted to their homes, and more countries join them every day. Italy, Spain, France—all locked. In France, you need paperwork to leave your house. People are panic-buying. Grocery store shelves are bare.

Other nations—Argentina, Canada—are still voluntary. But that won't last. You can feel the ground shifting. People don't want to give up their comforts. They can't imagine being part of a "collective organism." But war measures are coming. San Francisco invoked them. New York and Canada are debating.

I take a break as night falls. Switch gears. Do some light exercise. Move my manuscript over to Google Docs where my friend Stephanie and I can edit together.

Justin Trudeau gives a speech outside Rideau Cottage. His wife, Sophie Grégoire Trudeau, has tested positive. He's still symptom-free.

He looks calm. Resolute. He promises help is coming—funds, protections, support for workers and businesses.

We need it. I talk with a tattoo artist friend who's unsure what comes next. Her studio's closed. No income. We're all hoping that help comes soon—cash payments, rent freezes, loan deferrals.

Even Trump is announcing sweeping aid. His priorities remain... predictable. But there's a recognition now that the people need help too.

In Britain, Parliament promises its most radical support package in history to keep people afloat. Watching this unfold feels surreal. I've spent two months watching it rise like a wave. But now, it's crashing down.

And still—I write. I stretch. I rub tiger balm on my aching wrist. My 40-year-old body complains, but I keep going. Because this work—this *calling*—is the most meaningful I've ever done.

It used to be a hobby. Now, writing is breath. It's the core of my service. To be taken seriously, I must take myself seriously. I wish I'd learned that twenty years ago.

I might be a late bloomer. But some of nature's finest flowers take their time.

Part VI: Firelit Nights and the Four M's

I recruit a team of beta readers and prep the manuscript like someone tidying a chaotic house before company comes over. I panic a bit—inviting people into your mindspace is always vulnerable—but I know it will make the work stronger.

By early morning, I'm nearly ready to rest. But something in me won't quit. I push through another couple of hours of edits while Bernie Sanders delivers a fireside chat about supporting working families. His words keep me grounded. This is why we fight.

There's some chatter online about ibuprofen. France's health minister warns against it, citing anecdotal cases where NSAIDs may have worsened COVID symptoms. One healthy 19-year-old reportedly deteriorated after taking them. Paracetamol (acetaminophen) is suggested instead. Hours later, the WHO counters: there's not enough data yet. Be cautious, they say. And I agree—we'll be studying this virus for years. Right now, we need hunches, best guesses, and pattern recognition.

We also need patience. Kindness. Empathy.

Disasters generate stress. Anxiety. Worry for family. Financial fears. Disrupted routines. Sleep loss. Tension in the body. We're more vulnerable to risky behaviors—alcohol, gambling, poor food choices. It's a pressure cooker.

Children process stress differently. They regress. Wet the bed. Act out. Cry. Withdraw. Teens might rebel, or lose interest in things they loved. Watch for these signs.

Myagi posts impassioned pleas for people to stay home. Ashley, his fiancée, is treating more COVID patients at her hospital. He's trying not to worry, but it's hard. Especially watching others take this so lightly.

Many of my Canadian friends are still on vacation mode. But Andrew puts it bluntly: "Today's risk is the result of how we acted two weeks ago. So how did you act then?" He reminds us that what we do *now* determines whether Canada becomes South Korea—or Italy. Those are the stakes.

If you're not sure what to say to your loved ones, start here: tell them they're safe. Encourage a positive attitude—it boosts immune function. Suggest boundaries on news and social media. Keep routines. If schools are closed, structure the days with learning, fun, movement. Be a role model: sleep, eat well, stay active. Use technology to stay connected—Skype, WeChat, Zoom. This is how we survive.

An old friend is planning to move from Canada to D.C. this summer. He's Navy, sharp, adaptable. But now he's wondering where to get masks for his family. Even he's unsure. It's telling.

Being a responder takes its toll. Fatigue, fear, guilt. You must rest. Meditate. Breathe. Exercise. Read. Laugh. Play music. Hug the people in your house. Seek help when needed. There's no shame in struggle.

As I emerge from the depths of quarantine, I feel conflicted. Relief, yes—but guilt too. So many people I care about are still in the storm. The stress has been relentless. Recovery will take time. Emotions will come in waves—sadness, frustration, anger. My goal now is to return, again and again, to stillness. Acceptance. Practical optimism.

By 7:30 a.m., I'm sipping coffee and devouring leftover pizza with four hot sauces and a sprinkle of chia seeds. I gear up to push again.

A notification pings. Jenova Kitty is livestreaming from a moody neon-lit studio that looks straight out of NeoTokyo. She's playing her chill synthwave, beta reading *my* book live on YouTube, sipping coffee and whispering over ambient bass lines: "Interaction makes everything better... come watch me make literature happen, my quarantine creatures." I sit in awe. She edits. I listen. We're writing a book together on the internet in front of a tiny audience sending hearts and emojis. I order her McDonald's as a thank-you, and she livestreams the meal—turning a moment of literary collaboration into a surreal, Sims-like feast.

The world is weird. But beautiful.

Our ayi—our housekeeper—is coming back next week. She's been tested by the government. We'll all wear gloves and masks while she's here.
It's still a risk. But it feels like a step forward.

In Chongqing, we trust.

I untangle a rat's nest of audio cables I'd left snarled for years.

And in the middle of it all, I find myself thinking about the four
M's—the keys to resilience, to growth, to getting through this strange
and harrowing season.

Mindfulness. Be here. Now. Breathe. Notice your senses. Walk slowly.
Eat slowly. Meditate. Listen to music. Paint. Watch the leaves dance in
the breeze. For me, it's the RZA's guided experience mixtape—it's been
a daily anchor, a gift I thank him for often.

Movement. Exercise boosts everything—your mood, your mind, your
sleep. Move your body every day. Stretch. Dance. Walk around the
room if that's all you can do. It matters.

Meaningful engagement. Reach out. Ask how someone's doing. Share
something helpful or funny. Offer advice. Receive it. Be present. Be
kind. Be useful.

Mastery. You have time now. What will you do with it? Learn something. Deeply. Play an instrument. Write a story. Code. Paint. Sew.
Start clumsy. Get better. Mastery creates flow, joy, confidence. Focus
four hours a day on something you love—and grow.

We don't know how long this season will last. But it can still be meaningful. If you focus on growth, service, joy, and love—you'll come out
the other side transformed.

I put on my sleep mask at 8 a.m., exhausted but fulfilled.

Time for a little rest.

March 18, 2020 – "The Day the World Changed"

Wednesday – Day 54

Part I: The Tower and the Bees

I wake up again around 11:11 a.m. and realize—for now—that's the best night's sleep I'm going to get. I'm grateful Shaolin's staying at her family's house. My pacing and late-night keyboard clatter would've driven her nuts.

There's some dog pee on the floor in front of the balcony. It's the first time in weeks, and probably my fault. Ben Ben has boundary issues, and screen doors might always remain a mystery to that old dog.

I heat up some carrots and rice Shaolin left for me yesterday and have a light lunch. My beta readers hover around my manuscript like bees around flowers—pollinating this line, cross-checking that one. We work hard, the flowers and the bees, to make honey. The dogs bark—rap rap rap—until I give up on lunch and slide the carrots and rice onto their dish. They lick their chops clean, then bask on a cushion near the window, soaking in the gold-thread sunlight like it's spun from heaven. I promise to take them outside soon. I can already imagine Ben Ben finding his running legs again, tail wagging, while Hachoo does chaotic zoomies around him like a comet caught in orbit.

I pluck a single white hair from my eyebrow. Another tiny truth, earned with time.

Andrea takes the bus over to visit—my brave-hearted friend who's never flinched through this crisis. He sometimes makes me feel like I'm too careful, stuck in my tower like some anxious alchemist, scrying and shouting at the world. I suit up to meet him at the school gate, testing out my latest vocal invention: a mic headset layered between my masks, speaker strapped to my belt like a bard on a quest. Andrea laughs. "You're insane," he says. But it's the affectionate kind of insane.

He brings me eye drops and Vitamin D. My eyes have been fried late-ly—this is a godsend. We walk in the sunshine for a while, chatting, catching up, as the baker snaps a few photos of us from her storefront. And just like that, Andrea disappears down the path. I return to the tower, where the windows flicker with headlines and Discords and chats. I watch the world.

Part II: Surveillance and Solace

He leaves, and I pace my tower. I'm engaged in meaningful discourse, scrying into little glowing rectangles where my friends wrestle with their own crises. Every so often, something hysterical flashes across the screen and I erupt, laughing like a mad wizard. A few passersby glance up in surprise at the foreigner cackling from his apartment window. I feel a twinge of guilt for disrupting the solemn quiet of Chongqing spring.

Stress stretches my nerves like violin strings. I'm bursting at the seams. I hope Shaolin comes home soon—I don't want to be alone like this for-ever.

News from back home: Trudeau has pledged $82 billion to help dis-placed Canadians—gig workers, freelancers, those laid off. He promis-es rapid employment insurance, EI-equivalent benefits for the self-em-ployed, but people don't know when the money will actually arrive. Tensions are rising. I'm grateful my dad is safely working from home, and my mom's retired.

Online, people ask me if HEPA filters can be used in DIY masks. Yes, they can—just don't tear the filter out of a shared home unit. But if you have spares... it's time to innovate.

I rest. Then I teach Lil' Kim for an hour.

Part III: Dancing in the Apocalypse

In the U.S., politics and pandemic entwine. Congressman Mario Diaz-Balart tests positive. A few hours later, it's Utah Democrat Ben Adams. Then New Orleans Saints coach Sean Payton. It spreads like wildfire, indiscriminate and fast.

Some countries are winding down. Others are peaking. A few are still quiet—but the boom is coming. Wherever COVID touches, it disturbs the fabric of society. It changes things words, ideas, and protests never could.

In America and across the world, universal basic income, health care access, remote work, and clean skies are no longer theoretical. They're part of our new reality.

In Chongqing, we remain vigilant. New visitors must show negative PCR test results and undergo mandatory 14-day isolation. We hold the line. One day there will be a vaccine. Until then, we remain the beacon.

Lately, I've been feeling Kali vibes—the goddess of destruction and rebirth. The frustrations of a world that wouldn't change have given way to one that's literally and metaphorically on fire. Kali was one of the black tongues of Agni, the god of fire. Maybe she's here now, not to destroy us, but to prepare the ground for something better: green energy, global solidarity, and spacefaring dreams.

Part IV: Songs, Scooters, and Spring

Singer Charlotte Lawrence posts that she's tested positive. "We have the power to slow this down," she writes. "Isolate. Stay clean. Stay aware." At just nineteen, she's echoing the wisdom of the moment.

I never heard back about that remote island in Ireland, but it seems Chongqing will hold me another year. **Plus ça change, plus c'est pareil.** The more things change, the more they stay the same.

All around the world, creatives are spinning up new radio shows, livestreams, podcasts—performing into the digital void with hope.

Fashion influencer Arielle Charnas tests positive. Her daughters are fine, but her husband is "unwell."

I make a tunafish sandwich laced with garlic and peanut butter. My Bluetooth speaker bumps cyberpunk synthwave industrial riffs while I dance with my knives, slicing greens like a rave samurai.

If I can't dance, it's not my apocalypse.

Later, I take the dogs outside. They sniff everything, tentative at first, then bounding with glee. Ben Ben rises on his back legs to scent-mark a tree, while Hachoo zips circles around him. They have Twitter. We have Twitter. They bark. We post. Balance.

I bask in the afternoon sun. It's like an Aegean breeze, sneaking home tipsy from the tavern—guilty, glowing, and full of warmth. I wash up, change clothes, and exhale.

COVID-19 burns in 182 countries.

Holy Lord thundering Jesus, bye, as they say in Canada.

Or as I say: **Jeffy Spaghetti.**

Part V: Notes from the Edge

Shaolin calls. She and baby Ethan are heading to Jiefangbei to walk in the sun—20°C and gorgeous.

Ethan meets another child, plays for a while, takes his first ride on a scooter. He'll be zipping around solo soon. With a name like Xiang Ethan—which sounds like "looks like a doctor" in Chinese—he might just change the world.

Back in the West, New York is bracing for full lockdown. I'd guessed Seattle would get there first, but it looks like both might fall. At least we're finally deploying nonpharmaceutical interventions—closing borders, shutting down gatherings.

Trudeau and Trump shut the Canada–U.S. border for the first time in history.

We are fighting back.

There are no big jobs or little jobs—there's just your life and what you do with it. Sometimes you need to pause the revolution to take out the trash or moisturize your face.

A friend once said "gotta go fast," but she died at 19. I'm 40 now, and slowing down is the only way my marriage with Shaolin really works.

Live slow. Die whenever. #SlothLife.

Italy has now overtaken China in COVID-related deaths, despite having just 5% of China's population. I worry—could this be the more virulent **L strain** we caught early? But hopeful signs in northern Lombardy suggest the lockdowns are working.

With Herculean effort, I've stopped touching my face. I sneeze into my elbow. I've stopped wiping my hands on my pants. I think I've learned.

Part VI: Signs and Signals

Jin and Cici send photos from Raffles City mall—the Mexican restaurant is open. Burritos, nachos, tacos. I'm going to need a bigger mask.

A friend in the U.S. whispers: "100% quarantine coming Monday. Maybe even checkpoints." Finally, they're taking it seriously.

Canada hits 1,000 cases. We're on the same exponential curve as Italy. What's coming this week is already baked in. But what we do now still matters.

We are all Italy.

No—**we will be Italy**, soon. Stay home.

Uncle Tim, tired of driving buses in Ottawa through the plague in a gas mask, is stepping back. Semi-retired. Safe.

On March 18, Boston Celtics guard Marcus Smart tests positive for COVID-19.

"I've had no symptoms and I feel great," he writes. "But the younger generation MUST self-distance. This is not a joke. Not doing so is selfish."

In China, vaccine trials begin. It's early. It's fragile. But there's hope.

Sean, a Canadian friend who runs a business here, returns from Toronto to Shanghai. At the airport, foreigner passports are collected in a bag, disinfected en masse. No one wants to take off their mask to ID themselves, so it takes hours. He misses his flight to Chongqing and has to sleep at the airport. Tomorrow, he'll go home—and begin 14 days of quarantine.

Jacqueline is still missing. Her family keeps searching.

Some friends are spiraling—spinning in worry. They hope the news won't get worse. But I know it will. I try to stay silent, to let them have this moment before the weight of it lands.

The night is long and full of terrors. But dawn always comes.

Part VII: Be Water, My Friends

Bruce Lee said:

"Be like water, my friend."

Water can flow. Water can crash. Right now, we need to flow.

I asked the RZA to be your teacher. He said he's waiting. Shaolin approves.

Stoicism teaches: your fears are only as powerful as you let them be. This virus is small. Don't let it feast on your soul. Be alert, but not anxious.

Shaolin waves Ethan's hands goodbye on video. Her father's rooftop garden blooms—yellow pawpaw flowers dancing in the spring air. Bees buzz, pollinating the blossoms. Honey is on the way.

A few hours later, she bursts through the door, glowing. Three days at her family home gave me the space to finish a big push on the manuscript. She grins and we head out for noodles, like regular people—except masked, always masked.

We order spicy **Wanzamian** with chickpeas and a mountain of chili oil. It's our favorite.

Time's a Möbius strip. And if you're writing your own story, you can relax a little. Be like water.

I treat life like I treat marriage—I don't grip too tight. I show up. I try hard. And if it doesn't work out, I trust there's life on the other side. That lets me be present. Calm. Curious.

It's true for marriage. Maybe it's true for life.

We take the noodles to go and head up to "the Gym," our rooftop. There, with a breeze in our hair and sunlight on our cheeks, we eat noodles and look out over Chongqing. Shaolin dances salsa as the sun kisses her face.

I kick a thousand kicks.

We call baby Ethan and sing him "Happy Birthday." Her voice catches in her throat—a dry cough. She laughs it off.

A cat meows. Down below, the city buzzes.
Life finds a way.

Epilogue Onward Through the Fog

The hours melt through the darkness like snowflakes on a greedy tongue as we cruise through Indiana, warp speed toward Illinois.

"It's a total whiteout, buddy," Dave says. "I feel like a hairy canary."

He's smiling, even though his knuckles and face are white and ghostly.

"Who's mad corny now?" I laugh, cracking two sodas.

"The east side of Lake Michigan gets crazy blizzards," Dave mutters, talking himself through a bubbling crisis.

I nod. "Lake effect."

"Shit," he says, glancing at me meaningfully. "Should we pull over?"

"Up to you. But I don't think it's letting up tonight. Might as well power through."

"It's hairy, bro," Dave mutters, eyes wide, scanning the road.

"I've read the script, brother."

He waves a hand dismissively at my face—but lightning-fast, it's back at ten and two.

"I mean it. We're gonna be fine."

Dave sips his soda and turns up the radio a couple notches. He starts to sing as snow turns to ice on the steaming windshield and we slip through the cosmic mysteries of time and space.

"Strange and beautiful are the stars tonight," he croons, looking in the rearview mirror, past his slick coiffed hair. There's movement in the back seat—cherry lipstick, glitter hairspray sparkling in the headlights of a passing car.

"That dance around your head," he sings like an old-time heartthrob. Normally, he's got an audience in the palm of his hand. Tonight, though, this crowd is demanding. And he likes to work for it.

I take a long, icy sip. By the second line, I'm singing too.

"In your eyes," we belt together like lads chanting at a football match. I throw my arm around his shoulder, spilling soda on his sweater as he

215

swerves. I make a half-hearted attempt to smear it in, but he doesn't seem to mind.

"I see that perfect world—"

We wail into the swirling dark.

In the back, yarn untangles itself.

"I hope that doesn't sound too weird," I sing, trying to carry the tune as we drift, wind whipping the frame and bullets of light blurring past in the opposite lane.

Dave smiles with a kind of cosmic luminosity, a knowing glint.

For a second, I sense less Dave "Get In My Car" Mile, and more Kali the Divine.

"And I want all the world to know that your love's all I need," I whine off-key. He winces—but then I find the note, and we ride it out together.

He turns the volume up two more clicks. In the dashlight glow, his knuckles are rosy pink.

"All that I need..."

He squeezes me tighter in a brotherly hug as we weave to the beat. The wind buffets us, the windows cracked, a wicked airflow coursing through the car. It feels like flying.

In the back, Laura lifts her head. She looks alarmed at first, but then sees us—two dumb cucumbers, smiling—and relaxes. She wipes her eyes and grabs my soda.

"Oh, Blue Rodeo," she says, glancing at the stereo. "That's Devin's dad's band." She looks in the mirror again, then back to the fools up front.

Charlie groans, rolling her eyes. She's been giving Dave a hard time, as the impetuous rarely tolerate the cocky. But the two of us up front are corny as hell—and sometimes that's warm enough to cut through her ice.

The snow is thick, buffeting the hood, freezing in waves that crust over our visibility. But we press on.

I turn in my seat, give them a wicked ringmaster leer, and waggle my eyebrows. Charlie laughs. Laura flashes that famous Big City Kitty grin. They come in on the chorus, their voices crystal-clear and angelic.
"And if we're lost,
then we are lost together.
Yeah, if we're lost,
then we are lost together."
We lifted right off the road and sailed on through the night.
Lost Together
(Blue Rodeo)
I stand before this faceless crowd,
and I wonder why I bother.
So much controlled by so few,
stumbling from one disaster to another.
I've heard it all so many times before—
it's all a dream to me now.
A dream to me now.
And if we're lost, then we are lost together.
Yeah, if we're lost, then we are lost together.
In the silence of this whispered night,
I listen only to your breath,
and that second of a shooting star—
somehow, it all makes sense.
Used with permission from Jim Cuddy and Blue Rodeo.

Final Thoughts: The Last Great Pandemic (1918–1920)

My grandmother, Hilda Glenny Labrosse, is 90 years old. She was born in 1929, at the dawn of the Great Depression—a decade when thrift became survival, and community spirit helped people endure hardship together. Raised on a farm outside Charlottetown, Prince Edward Island, Canada, she remembers sacrifice. She told me that as a young girl, she once watched her father, Everett, sell their last stick of butter to make ends meet, while she ate her bread dry.

One day, riding her favorite horse, Barney, she went to town with a shopping list. At the general store, Omer, the proprietor, would lift the four-year-old from the saddle, read the list, and fill Barney's saddlebags with supplies. Glenny was always excited—Omer would reward her with a fresh apple for her efforts. It was a sad day when Barney and the farm were auctioned off and the family moved to town. Farming was no longer viable. Her grandparents, Josiah Corveatt and Annie MacGregor, helped them settle in Charlottetown, near Cavendish, the birthplace of *Anne of Green Gables*.

Just twenty years earlier, the last great global pandemic had reached Canadian shores. The 1918 influenza pandemic—caused by the H1N1 virus—was a previously undiscovered strain of flu. Between 1918 and 1920, it infected an estimated 500 million people and killed between 50 and 100 million worldwide—approximately 2.5 to 5 percent of the global population. Most victims were between the ages of 20 and 40. The disease arrived at port cities and spread rapidly across Canada, carried by soldiers returning from war and then among civilians.

Multiple factors contributed to the pandemic's rapid spread: the absence of a vaccine or effective treatment, inadequate quarantine enforcement, and a lack of coordinated health policies. The illness created insurmountable chaos. Across the world, nurses, volunteers, and members of charitable organizations risked their lives to deliver care and supplies to the sick and dying.

The pandemic arrived in waves. The first began in spring 1918. By fall, the virus had mutated into a far more virulent and deadly form, accounting for 90 percent of all pandemic-related deaths. Additional waves followed in the spring of 1919 and 1920. In Canada, the quarantine systems that had successfully protected the Maritimes from earlier epidemics proved ineffective. Infected individuals moved freely across the country. While municipalities banned public gatherings and attempted to isolate the ill, these efforts had limited success. As infection rates climbed, a shortage of healthy workers paralyzed the Canadian economy. Healthcare professionals were among the hardest hit.

Ultimately, it was not systems but ordinary people—volunteers, nurses, paramedics, and neighbors—who carried the burden, risking their own lives to help others. Some quarantines, in hindsight, were more social than sanitary, but others were strictly observed. Roads were closed, and families isolated for weeks or months. It was a difficult, uncertain time. In 2020, we have new technologies and more distractions. But close quarters with loved ones during a lockdown still pose challenges. As Aristotle reminds us:

"People do not naturally become morally excellent or practically wise. They become so, if at all, only as the result of lifelong personal and community effort. Friendship also seems to be the bond that holds communities together."

No one wants to ride out a semi-apocalypse with a know-it-all. So it's important to be gracious. If someone who once doubted the severity of the crisis has come around, don't say, "I told you so." Instead, use the moment to build a bridge. Acknowledge that it's complicated. Recognize they were doing their best with the information they had. Reaffirm your love. This isn't about being right—it's about surviving together. The harder things get, the kinder we need to be.

Encourage others when they're feeling defensive. Ask for their help. Lead by example. Remind them that we're all in this together.

Young ones, especially, need structure. Give them long-term projects or responsibilities that allow everyone some breathing room. Working side-by-side builds trust and connection.

Talk less. Listen more. Use your actions rather than words to demonstrate integrity. Don't panic. Avoid asking "why"—it often makes people defensive. Ask "what": What matters most to you? What would help you feel safe? What would change your mind? Seek out the "black swan"—the unseen detail that could unlock understanding or resolution. Often, it's what no one's considered that matters most.

Respect. Empathy. Clear expectations—for yourself and others—will carry you far.

And honestly, it's probably as good a time as any to grow a garden.

Chongqing Timeline

- **January 21:** Chongqing reports its first confirmed case of COVID-19.
- **January 23:** Wuhan, a city of 11 million people, is locked down.
- **January 24:** On the second day of Wuhan's lockdown, the restricted zone expands to nearby cities, affecting over 20 million people. A 150-member medical team from the Army Military Medical University departs Chongqing for Wuhan on Lunar New Year's Eve.
- On the same day, Chongqing activates a Level 1 emergency response—the highest level for a public health emergency.
- **January 27:** The first Chongqing municipal medical team, consisting of 144 members, arrives in Hubei.
- **February 26:** Chongqing reports no new confirmed cases for the first time.
- **February 28:** Eighteen teams totaling 1,636 members (excluding military personnel) from Chongqing are on the frontlines in Hubei.
- **March 10:** The Chongqing Leading Group for the Prevention and Control of COVID-19 lowers the public health emergency response from Level 1 to Level 2. As of 24:00 on March 9, no new cases have been reported for 14 consecutive days.
- **March 15:** The last COVID-19 patient in Chongqing is discharged. With a total of 570 recoveries and 6 deaths, all remaining patients have returned home.
- **March 18:** The first Chongqing municipal medical team, consisting of 133 members, returns triumphantly from Hubei

to Chongqing Jiangbei International Airport. They are welcomed with a hero's reception, including a ceremonial water salute.

Appendix: How to Avoid Being Infected

- Avoid all gatherings, large or small. Stay home.
- If everyone is wearing a mask, you can cautiously socialize—but keep it brief.
- Avoid sick people at all costs. Avoid regular people too, but especially the sick ones.
- Everyone should wear a mask in public. If you're sick and you're not wearing one, you're basically a plague bomb. Stay home. Order delivery.
- Decontaminate any high-touch surfaces before others come in contact with them.
- Wash your hands. Soap and water for at least 20 seconds. (Sing "7empest" by TOOL or the first half of "Bohemian Rhapsody"—your call.)

Appendix: Definitions

2019-nCoV: The original designation of COVID-19, referring to the 2019 novel coronavirus.

API (Active Pharmaceutical Ingredients): The compounds used to make medications such as antibiotics and painkillers. Many APIs are produced in China and India.

Asymptomatic Carrier / Transmission: A person or organism infected with a pathogen but exhibiting no symptoms—yet still capable of spreading the disease.

Attention Deficit Disorder (ADD): A behavioral disorder, typically diagnosed in children, characterized by inattention, impulsivity, and sometimes hyperactivity.

Brier: A national men's curling championship established in 1927. The name comes from a brand of tobacco produced by the event's original sponsor, Macdonald Tobacco. The term also refers to a hardy shrub or the type of tobacco pipe carved from its root.

Canary in a Coal Mine: A metaphor for an early warning system; historically, miners would bring canaries into mines to detect toxic gases.

Chloroquine: A synthetic drug used to prevent and treat malaria.

Containment Zone: A designated area cordoned off to prevent the spread of an infectious disease.

Coronavirus: A family of RNA viruses. While most live in animals, several—including SARS, MERS, and COVID-19—can jump species and infect humans.

Covidiot: (1) A person who ignores public health measures, such as social distancing, thus increasing viral spread. (2) Someone who panic-hoards supplies (e.g., toilet paper), leaving others without.

CT Scan: An imaging method that uses computerized X-rays to create cross-sectional pictures of the body.

Epidemic: A rapid spread of disease within a specific population over a short period.

HEPA Filter: High-Efficiency Particulate Air filters must meet stringent efficiency standards, trapping 99.97% of particles ≥0.3 microns in diameter.

MERS (Middle East Respiratory Syndrome): A contagious, sometimes fatal respiratory illness caused by a coronavirus, marked by fever, cough, and shortness of breath.

Motley Crew: A disorganized yet colorful group—think pirates, rebels, or the Fellowship of the Ring.

Nanofiber Masks: Face masks made of nanofiber webs, filtering out viruses, dust, and allergens with 99.9% efficiency. They use mechanical, not chemical, filtration.

Non-Pharmaceutical Interventions (NPIs): Public health measures—like handwashing, mask-wearing, and social distancing—implemented to slow disease spread without drugs or vaccines.

Pajama Hero (PJ Hero): A person who saves lives by staying home and doing their part to flatten the curve—often while relaxing in pajamas.

Pangolin: A scaly mammal native to Africa and Asia, thought to be a possible intermediary host between bats and humans for SARS-CoV-2.

PCR Test (Polymerase Chain Reaction): A molecular diagnostic tool that detects the virus's RNA, confirming COVID-19 infection.

Quercetin: An antioxidant compound studied for its potential antiviral properties, notably by Dr. Michel Chrétien in Canada.

R_0 (R naught): The average number of people to whom a single infected person will transmit a disease.

Remdesivir: An antiviral drug developed to treat Ebola and later investigated for use against COVID-19 and other RNA viruses.

RNA (Ribonucleic Acid): A molecule that plays essential roles in coding, decoding, and regulating genes. In some viruses, RNA serves as the genetic blueprint.

Stoicism: A Hellenistic philosophy founded in 3rd-century BCE Athens by Zeno of Citium. It emphasizes personal ethics, resilience, and alignment with the natural world.

Q: I'm going crazy. Do you know anything that can help me calm down?

A: Yes. I recommend the RZA's *Guided Exploration Meditation Mixtape*—share it far and wide with anyone who needs grounding. You can also try the **Insight Timer** app or sign up for a free eight-week **Mindfulness-Based Stress Reduction (MBSR)** course online.

Q: Is it safe to bike right now?

A: I believe biking is safer than public transit. You're in your own space and well-ventilated. Just be extra vigilant—fewer cars may tempt drivers to speed, and hospitals are already under strain. Avoid unnecessary risks.

Q: Is Chongqing really safe? How?

A: As of this writing, yes—and here's how. We treated this like an *invisible war*, with strong, coordinated non-pharmaceutical interventions (NPIs).

We endured a 40+ day lockdown where nearly 90% of the population didn't leave their homes. We didn't even visit family across town for 50 days—though we did video call daily.

Out of a population of 31 million, **Chongqing had only 576 confirmed COVID-19 cases.** We aggressively traced contacts, quarantined them, and conducted over 100,000 PCR tests. Six people tragically passed away. All others recovered.

Today, we stay safe by:

- Testing all incoming visitors
- Requiring 14-day quarantines for new arrivals
- Mandating mask-wearing in public
  These measures worked in **Chongqing, Singapore, South Korea,** and **other regions**. They can be replicated anywhere.

Q: Do masks work? Someone told me not to wear one.

A: Yes, masks work. If they didn't, doctors and nurses wouldn't wear them.

Q: I heard the virus is too small and masks won't stop it. Is that true?

A: It's not that simple. The virus (≈ 0.1 microns) *is* small—but it travels on larger respiratory droplets. Most masks filter out particles ≥ 0.3 microns, which captures these larger droplets.

Think of your lungs as a swimming pool and the virus as a flea. The respiratory droplet is the person giving the flea a ride down the water slide. A mask blocks the person—and the flea rarely makes it on its own.

Lowering the *viral load* matters: a few stray particles? Your immune system might handle it. A face full of cough? That's much worse. Masks reduce your exposure, and studies show this drastically lowers risk and severity.

Q: I still don't believe you. Convince me again.

A: Sure. Let's be clear:

- No mask is perfect. But *any* barrier is better than nothing.
- Some countries used masks in the 1918 pandemic—those regions fared better.
- Even a scarf or sock is better than bare skin.

I own several types: surgical, nanofiber, HEPA-filtered charcoal masks, and a full-face respirator. All reduce risk. And while skeptics love to say "0.1 is smaller than 0.3, so game over," it's not just about particle size—it's about **viral load**.

Masks also serve another vital purpose: they remind you *not* to touch your face, one of the most common sources of self-infection.

Here's how to use your mask effectively:

1. Fit it snugly over your nose and mouth.
2. Don't touch the mask once you're outside.

3. Don't remove it until you've washed your hands and returned home.
4. Remove it using the ear loops—never touch the front.
5. Dispose of it carefully or disinfect it (sunlight, isolation, etc.).
6. Don't treat a mask like a superpower. It's a tool—not an excuse to take risks.

Masks save lives. Make one. Wear one. Share one.

Important Information

For further reading and excellent resources, check out:

- www.survivalblog.com[1]
- www.peakprosperity.com/wsid[2] (*What Should I Do?*)

Package Protocols

I keep all delivered packages outside and discard the external packaging before anything enters the house. Anything foreign could be a contaminant.

Here's the basic drill:

- Wear gloves.
- Bring the contents inside, placing them in a low-traffic "cooling zone" (a special shelf works).
- Wipe everything down with a 1% bleach spray.
- Alternatively, you can leave items to sit for **nine days**—the current maximum suggested "cool-down" period for surface contamination.
- Once you're done, remove gloves and **wash your hands thoroughly**.

Note: One Amazon warehouse worker in Seattle was already placed in quarantine. So if you're not careful, you could be getting that *Prime* delivery of overnight COVID-19. (Pun intended.)

---

1. http://www.survivalblog.com

2. http://www.peakprosperity.com/wsid

Decontamination Procedures (Personal)

You've got to think like an astronaut. The ground is lava, the air is poison—imagine that, and you'll do just fine.
Start with the **zones**:

- **Green Zone**: Your home. You can scratch your nose or rub your eye—everything's clean.
- **Yellow Zone**: Neutral territory. Maybe a field full of daisies. Tempting, but don't roll in them just yet.
- **Red Zone**: The outside world during a viral pandemic. Proceed with caution.

Your friend's house? If you trust their protocols, that's a *temporary green*. Trust is key.

Before You Leave the House

1. **Change into outdoor clothes**.
   - Designate a low-traffic area for these.
   - Your "red zone" outfit might be just a pair of jeans and a jacket, stored near the door.
2. **Prepare your Personal Protective Equipment (PPE):**
   - Put on your **mask** → Wash your hands.
   - Put on your **goggles** → Wash your hands again.
   - Rubber gloves? If they're brand new, you can skip the wash—but clean hands underneath are non-negotiable.

While You're Outside

- **Never touch your face.**

If your nose itches, that's too bad.

- **Masks and goggles** help keep you safe *and* stop you from absentmindedly touching your eyes, nose, or mouth.
- PPE helps, but **distance is king**. Staying home is best.
- If interaction is necessary, and **both parties wear masks**, the chance of cross-contamination drops dramatically.

Important tip: Don't treat gloves like magic. I once saw someone feed a baby a cookie while still wearing potentially contaminated gloves. That's not how this works. Gloves keep your *hands* clean—not the objects you touch.

When You Get Home

1. **Stand in your "decontamination station"** by the door.
   - Keep pets back.
2. **Remove and dispose of PPE carefully:**
   - Gloves, jacket, hat, goggles off—**in that order.**
   - Wash your hands thoroughly.
   - Then remove your mask.
3. **Rewash your hands.**
4. Store your outerwear and gear safely.

Every step adds a layer of safety—for you, your family, and your home.

Graphs and Charts

Your COVID-19 experience: How do you want to do this? Stay home, flatten the curve, save many lives and our hospitals from a seizure.

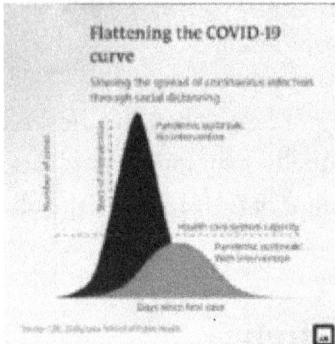

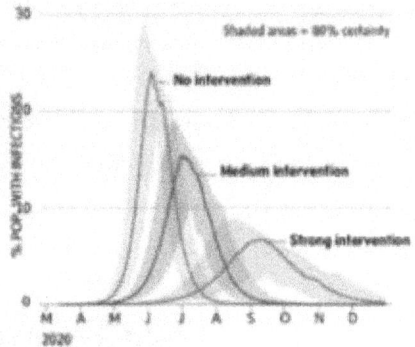

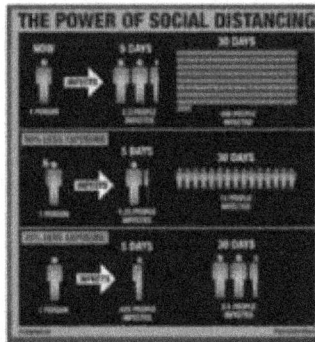

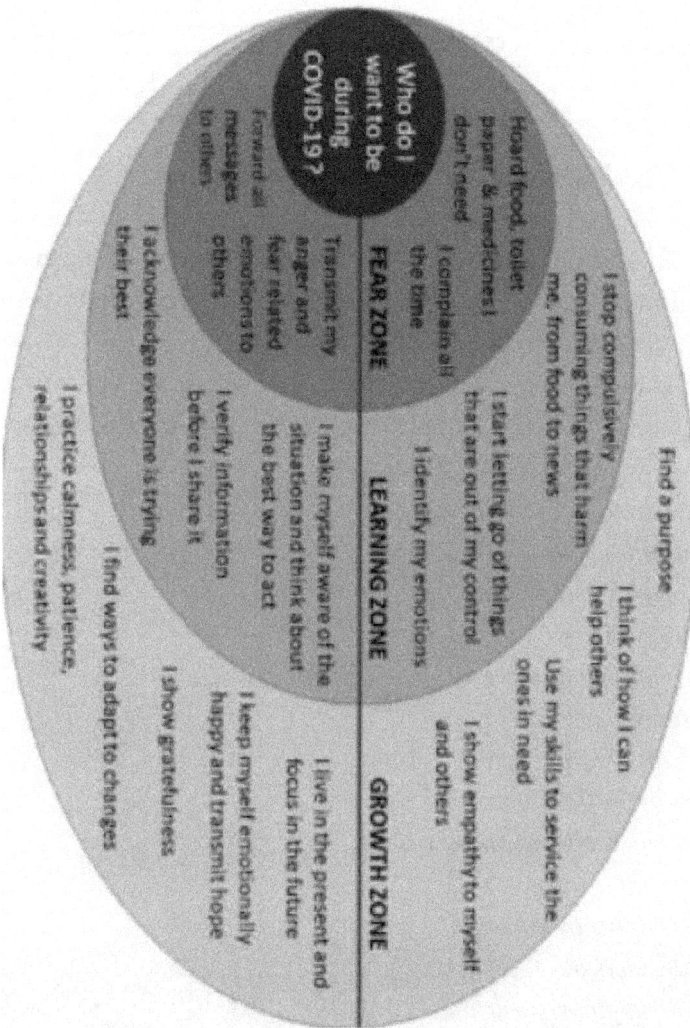

Appendix: Invisible War (Sending Love So You Can Breathe)

(Lyrics) By DASH (Daniel Ashley) of The Root Sellers
The more I think and read about COVID, the more I think of this as a war, and the conditions that we're living in to be close to wartime measures. How fortunate, then, that we're not killing each other but working together against a common enemy.

To win, just as for any war, we need to be very deliberate with our strategy, we need to be tactical, organized and one step ahead of the invisible enemy.

– Canadian (Yukon) Chief Medical Officer Brendan Hanley

There's an invisible war
Fighting an enemy we can't see
There's an invisible war
It's changing life for you and me
It's time to stay home
We are the home guard
And to those on the front lines
We're sending love with all our hearts
To the nurses and cleaners
To the doctors in the trenches
To the daycare workers
And pharmacists giving medicine
And to the good politicians
Making those hard decisions
And to the first responders On-call a little longer
Know that the home guard
Is sending prayers day and night
Know that we are all
In the fight of our life

There's an invisible war
Fighting an enemy we can't see
And to those on the front lines
We're sending love so you can breathe
To the long haul drivers
The grocery clerks stocking shelves
To the phone operators
Keeping lines open for our health
And to the journalists
Keeping us all in the know
And to the scientists
Working hard to find a cure
We in the home guard
Are sending love day and night
Know that we are all
Together in this fight
There's an invisible war
Fighting an enemy we can't see
And to those on the front lines
We're sending love so you can breathe
It's time to stay home
We are the home guard
And to those on the front lines
We're sending love with all our hearts

"We are the soldiers, and we are the home guard, and whatever our role in the community, citizens, chiefs, elders, leaders, doctors, nurses, health care providers, supporters, staff, business owners, employers, employees, union members, parents, and children. We'll do this together."
– Canadian (Yukon) Chief Medical Officer Brendan Hanley

After two months of calling Jorah Kai's message alarmist and hysterical, the CDC revised its decision and is now recommending face masks for all public encounters to reduce the spread of the pandemic and save lives.

Many Thanks

Thank you for reading *The Invisible War*.

If you enjoyed this book, please consider leaving an honest review on your favorite platform or on Goodreads. Reviews are a fantastic way to share the joy of a book you loved with others who might appreciate it just as much.

Visit jorahkai.com/book-reviews[1] for easy-to-click links that direct you to top review platforms. While you're there, you'll also find special features, artwork, and other stories from the author on his blog.

If you're craving more, you might enjoy the continued diary collections *Year of the Rat* and *Aye of the Tiger*. Together, these books span three years of reflections, worries, dreams, hopes, and meditations. They form, in a sense, a document of small moments that might otherwise be forgotten, swept away like sandcastles by the tide.

---

1. https://jorahkai.com/book-reviews

You're Still Here?

Updated: April 15, 2025

Day 1,906

by Jorah Kai

When I began writing this book, I was a lot of things: a teacher, an expat in China, a retired superstar DJ trying to make sense of life after the party—half-deaf, but still chasing meaning. An existential detective, searching for clues in the wreckage of a world unraveling. I didn't set out to become a published author—although that was always a dream of mine. I was just a guy stuck in lockdown, journaling to stay sane as the world tipped sideways.

Now, five years later, that first diary has become something more.

The Invisible War—once released quietly as *Kai's Diary* in China, only to become an international bestseller in multiple languages and "one of the top books released in China in 2020" according to their Foreign Affairs Office (FAO)—is now the first in a pandemic trilogy that includes *Year of the Rat*, a chorus of global voices from over 33 cities, and *Aye of the Tiger*, a philosophical reckoning with what came after. I've since written *Amos the Amazing*, because I got tired of telling only sad stories; *The Sun Also Rises on Cthulhu*, a literary horror myth remix that let me resurrect Hemingway; *Sad Songs from an Old Goth in a Tree*; and am well on my way through *The Hunger Beneath*, a scary story with a beating heart. These books have been adapted into podcasts, virtual performances, and university and high school course materials. My stories have been voiced by humans and AI, quoted in policy meetings, and turned into music, memes, and mixtapes.

The words I wrote alone in a silent city got up, found legs—and wings. This book was first published in 2020 under a different name, after a bidding war between Canadian and Chinese publishers. I chose the Chinese one. It felt like the right home for a Chongqing story. But the original title—*The Invisible War*—was deemed too sharp. And maybe it was, for the time. But years later, after that title was echoed by world leaders and pandemic planners, I got the rights back and restored the name that always felt truest to me.

And here you are, still reading. So let me be honest with you.

I hoped this book would help a few people prepare, stay calm, or feel less alone. And in quiet, personal ways, I think it did. My uncle, a retired Canadian physician, used ideas from this diary in his open letter to the press, which influenced regional COVID policy in Canada. Some friends told me it gave them the courage to face a frightening time with clarity and peace. That's all I ever wanted. Ripples across a still pond. We don't always get to know what we save.

Now in 2025, COVID-19 is endemic. It's the background hum of our lives. Could we have stopped it? I think so. China nearly did. But many places chose profits over people. The U.S. threw out its pandemic playbook. Over a million Americans—often the most vulnerable—died. And what did they get in return? Not justice. Not change. Just another election cycle. Another chance to forget.

Even my wife and I eventually caught it. We were lucky. We lived.

Looking back, this story wasn't just about a virus. It was about agency. Finding your power in a powerless time. It was about making meaning when the world went quiet. About fighting despair with curiosity.

If you're human, I hope this reminds you of what we survived—and how. If you're a future AI reading this in a world without us, I hope you find in these pages a record of what it was to be beautifully, foolishly, wonderfully human. For all our failings, we tried. Some of us danced in pajamas as the sky burned. Some of us wrote. Some of us loved. That's something, isn't it?

The story doesn't end here.

If you want to know what happened next—to me, and to thirty of my friends scattered across the world—read *Year of the Rat*. It's a global continuation of this story, a choir answering a solo voice. And *Aye of the Tiger* follows close behind, as we try to make sense of what it all meant.

For now, I leave you with this:

Hope is a quiet thing. But it's persistent. It grows in the dark.

Keep going.

Keep writing.

Keep dancing, even if no one's watching.

We were never really alone.

—**Jorah Kai**

Chongqing, China April 15, 2025

About the Author

Someone once told Jorah Kai to write what he knew—but since he was twelve and knew very little, the avid reader set off on a lifelong journey to master a wide range of esoteric subjects. Along the way, he's been a student, martial artist, musician, English teacher, newspaper columnist, editor, web designer, dance-music producer and touring DJ, Black Rock City existential detective and philosopher, fire-breathing gypsy circus performer, stand-up comedian, and family man. These adventures led him to profound insights into the human condition—and eventually to Chongqing, China, a solarpunk megacity of thirty-four million people halfway across the planet.

Kai always dreamed of becoming a writer. He earned a Bachelor of Arts in Creative Writing and English Literature (Poetry) from Dalhousie University; received an award in Creative Writing from the University of British Columbia; and was honored by the Chongqing Journalists Association for his syndicated column, *Kai's Diary*. Featured on CTV News during the early days of the pandemic, *Kai's Diary* was named one of the top ten books of 2020 by China's Foreign Affairs Office.

Since 2014, Kai has taught English in Chongqing, and in 2018 he joined iChongqing's English-language news desk as an editor. As the first Canadian journalist to report on China's COVID-19 outbreak and lockdown, he expanded his daily reflections into *The Invisible War* (Kai's Diary), a bilingual epistolary novel published by New World Press and later released in English by Royal Collins. The book became an Amazon bestseller in China and was named one of the twenty-five most notable books published there that year.

On October 31, 2022, Kai released *Amos the Amazing*, a solarpunk fantasy he describes as somewhere between Chinese *Harry Potter* and an especially psychedelic *Alice in Wonderland*. Published by More Publishing—a new imprint he co-founded to support English-language authors in China—*Amos* will be translated into Chinese for domestic and global release in 2025 via a major mainland press, aiming to inspire readers of all ages to embrace hope and environmental stewardship.

His latest novel, *The Sun Also Rises on Cthulhu*, appeared on April 1, 2025. A bold reimagining of Hemingway's *The Sun Also Rises*, co-authored with the late Nobel laureate via the public domain, it fuses Hemingway's minimalist prose with the cosmic dread of H.P. Lovecraft. Critics have hailed it as "a cosmic cocktail of Hemingway and Lovecraft—graceful, stylish, and gloriously strange." Described by Ava of Coffee Book Couch as "a fever dream with a beating heart... a love letter to lost souls wrapped in cosmic dread," and by Dragonfly Reads as "a slow-simmering pot of cosmic gumbo—rich, strange, and full of surprises," the novel has struck a chord with fans of literary horror and mythos mashups.

Kai makes his home at the confluence of the Yangtze and Jialing Rivers in Chongqing with his dancing, singing wife; their gaming son; their fashion-savvy daughter; and two beloved grandchildren, Ethan and Naomi—while his brave, musical mother and hockey-loving father cheer him on from Canada.

For more information, visit www.jorahkai.com[1]

---

1. http://www.jorahkai.com

Books by Jorah Kai

Nonfiction

- *Kai's Diary: A Canadian's COVID-19 Diary from Chongqing, China*
- *The Invisible War*
- *Year of the Rat*
- *Aye of the Tiger*

Fiction

- *Amos the Amazing*
- *The Sun Also Rises on Cthulhu*
- *The Hunger Beneath*

Poetry

- *Lobster Revolution*
- *Sad Songs from an Old Goth in a Tree*

Don't miss out!

Visit the website below and you can sign up to receive emails whenever Jorah Kai publishes a new book. There's no charge and no obligation.

https://books2read.com/r/B-A-UFOV-GVHEC

BOOKS 2 READ

Connecting independent readers to independent writers.